Makers of Modern Science

NIELS BOHR
GENTLE GENIUS OF
DENMARK

Ray Spangenburg and Diane K. Moser

AN INFOBASE HOLDINGS COMPANY

NIELS BOHR: GENTLE GENIUS OF DENMARK

Facts On File, Inc.
460 Park Avenue South
New York NY 10016

Library of Congress Cataloging-in-Publication Data

Spangenburg, Ray, 1939–
    Niels Bohr: gentle genius of Denmark/by Ray Spangenburg and
Diane K. Moser.
        p. cm.—(Makers of modern science)
    Includes bibliographical references and index.
    ISBN 0-8160-2938-5 (alk. paper)
    1. Bohr, Niels Henrik David, 1885–1962. 2. Physics—History.
    3. Physicists—Denmark—Biography. I. Moser, Diane, 1944–.
    II. Title. III. Series.
    QC16.B63S63 1995
    530'.092—dc20
    [B]   94-23815

Text design by Ron Monteleone
Illustrations on pages 28 and 66 by Marc Greene
Cover design by Catherine Hyman
Printed in the United States of America

RRD FOF 10 9 8 7 6 5 4 3 2 1

This book is printed on acid-free paper.

*In memory of Margaret L. Notson Moser
who always encouraged
a love of ideas, knowledge, and understanding*

ACKNOWLEDGMENTS

We wish to thank the Niels Bohr Archives in Copenhagen for generous assistance in gathering materials for this book. We would also like to thank Shawn Carlson, of Lawrence Berkeley Laboratories, for reading the manuscript and making many helpful suggestions. A special thanks, as well, to our editors, James Warren and Jeffrey Golick, and the rest of the team at Facts On File.

CONTENTS

PROLOGUE

The cold late-winter wind of 1948 swept through the sparse woods behind the Institute for Advanced Study a few miles outside of Princeton, New Jersey. On the institute grounds that late afternoon only a few solitary walkers could be seen scurrying from building to building. A battered old car backfired a short distance away as it pulled up in front of Fuld Hall and discharged a passenger, a tall young mathematics professor with a hurried air. With his coat collar turned up against the wind, he rushed inside to the red-brick warmth of Fuld Hall, carrying an overflowing armload of books and papers from the nearby Princeton University library.

The Institute for Advanced Study, founded in 1930, had been established, as the founders' letter explained, to foster "the pursuit of advanced learning and exploration in fields of pure science and high scholarship to the utmost degree that the facilities of the institution and the ability of the faculty and students will permit," with the emphasis, from the beginning, on the word "advanced."

Although the founders had envisioned a kind of supergraduate school, where the brightest faculty minds would be free to pursue their own studies while teaching only a hand-picked group of graduate students, in reality the institute never enrolled a single graduate student. Its "students," instead, were its own brilliant and gifted members, an international community, all holders of the highest degrees and honors, the best scientific minds in the world. Freed from nonacademic distractions and responsibilities, the institute's members could concentrate solely on their own work, and they eagerly listened to and taught not undergraduate or graduate students, but one another, exchanging ideas with the best of their peers.

Not surprisingly, the institute drew many of the world's finest scientific and scholarly minds, both for long-term stays and shorter visits. In 1947 the "father of the Atomic Bomb," J. Robert Oppenheimer, had become the institute's third director. Under his leadership, as under the two previous directors, the big comfortable offices in Fuld Hall continued to be a thriving beehive of esoteric and often bewildering intellectual activity.

On that particular late-winter afternoon in 1948, perhaps its most prestigious resident member was Albert Einstein.

The young mathematics professor just in from the cold hurried through the office corridors with his armload of books and his hair still mussed from the wind, glancing in the open door of Einstein's cluttered office. He had promised to drop off some papers, but no one was in. He would drop them off later.

On another floor, in another part of the building, Einstein was also on the minds of two other men. The younger of the two, Abraham Pais, a young scientist wearing horn-rimmed glasses, sat at his desk, watching. The other, an older man with a large stocky frame and the drooping face of a big, friendly bulldog, paced the room. The older man muttered softly, his thick Danish accent making his words almost unintelligible. He was visibly upset. He had jammed the pipe he usually smoked into his pocket, and he seemed to be struggling even more than usual for the right words. Finding them hardly seemed a relief.

"I am . . . I am . . . sick of myself," he muttered finally, giving the younger man a frustrated look before stalking back to stare out the window at the late-afternoon sky.

The younger man watched sympathetically. It wasn't the first time that he had witnessed Niels Bohr's frustration after one of his long and difficult discussions with Einstein. It might not have been as hard if the two men hadn't cared so much for each other.

"I am as much in love with him as you are," Einstein had once written about Bohr to a friend. In another letter, to Bohr after their first meeting, he wrote, "Not often in life has a human being caused me such joy by his mere presence as you did."

But, despite their affection, their heated and complex intellectual debate had gone on now for over two decades and neither had managed to change the mind of the other. Each was completely certain that he was right. And with that certainty came constant frustration over their mutual inability to convince the other.

A bubbling cauldron of intellectual energy, Niels Bohr didn't take frustration easily. Pais remembered Bohr's arrival at the institute only a few weeks before. Bohr had taken a ship from his home in Denmark. Finding himself aboard without anyone to discuss his ideas with, he had practically exploded from the frustration. Niels Bohr was a man who needed to talk. The first people he saw in the institute's corridors had been Pais and Wolfgang Pauli, and he had quickly hustled them into an office, and after silencing the talkative Pauli, proceeded to unravel all the thoughts collected during his ocean voyage. Pais later remembered wishing that there had been a stenographer handy. Bohr's conversation was a veritable history book on the evolution of ideas about the atom and its structure, ideas that had forever changed most physicists' views about reality.

Actually, "conversation" was a misleading word. It had been more of a mumbled, occasionally hesitant, but always brilliant monologue,

The Institute for Advanced Study in Princeton, N.J. (Courtesy of the Archives, Institute for Advanced Study, Princeton, N.J.)

with even the usually argumentative Pauli sitting quietly and listening, spellbound.

Now, though, Bohr stood silently at the window. Perhaps he hoped to catch a glimpse of Einstein on the grounds. Einstein, who it seemed had haunted his mind ever since that first meeting in Berlin in 1919. Einstein, who would never agree with the ideas that Bohr felt so terribly strongly about, who would never see the world in the way that Bohr saw it. Einstein, who would continue to go his lonely and obstinate way.

The big Dane shrugged his bulky and still-athletic shoulders. He would never win Einstein over. Never. Never! And yet . . . And yet . . .

The young mathematics professor dumped his armload of books onto his desk. He was running late and was in a hurry to wrap up his errands and get home for dinner. The wind was blowing harder and was rattling the window of his office. A storm was coming and he didn't like his wife to drive in bad weather. If he could get back downstairs, he could catch her in time before she parked the car and came looking for him in the building.

He caught a movement in the corner of his eye. Glancing up, he saw the familiar shaggy-haired figure walk past the open door. He started to call out, checking his desk to make certain that he still had the papers Einstein had asked him to pick up. When he looked up again, he could no longer see the baggy gray sweater. He debated with himself briefly. Did you chase Albert Einstein? He moved to the door and looked out into the corridor, but he had acted too slowly. The shaggy figure had vanished around a corner.

Outside the institute, the skies had darkened and the wind had picked up. Lights were going on in many of the offices. In a window on one of the upper floors, Niels Bohr stood silently staring out at the wind-swept grounds.

A dozen windows away, another light flicked on. Packing the pipe that his doctors had forbidden him to smoke with tobacco he had "borrowed" from one of the other offices, Albert Einstein moved to the window and stared out.

On the grounds below a horn honked. The car door opened for the hurried young professor—and Einstein and Bohr returned to their thoughts. They would still be puzzling for hours to come. Neither would sleep well that night. And tomorrow they would try again.

What was this argument all about, an argument that engaged two of the most brilliant scientists of the twentieth century in a heated debate that Robert Oppenheimer once called "the richest and deepest dialogue since Plato's *Parmenides*"? A debate about nothing less than the nature of reality. And who was Niels Bohr, this tenacious but gentle bulldog of a man who for over two decades engaged in this great intellectual debate with Albert Einstein?

PROLOGUE NOTES

p. xi   "the pursuit of advanced learning . . ." Brochure published by the Institute for Advanced Study, Princeton, New Jersey, 1954, p. 1.

p. xii   "I am . . . sick of myself." Quoted in *Niels Bohr: A Centenary Volume,* edited by A. P. French and P. J. Kennedy, p. 248.

p. xii   "I am as much in love . . ." Quoted by Richard Rhodes, *The Making of the Atomic Bomb,* p. 54.

p. xii   "Not often in life . . ." Quoted in Rhodes, p. 54.

p. xv   "The richest and deepest dialogue . . ." As quoted by Ruth Moore, *Niels Bohr: The Man, His Science, and the World They Changed,* p. 397.

1

DENMARK: DAYS OF SOCCER AND SCIENCE 1885–1911

It was an important game between the world-famous Danish soccer team and the top-notch team from Germany. The Danes were attacking strongly on the German end of the field, when the direction of play turned suddenly toward the Danish team's goal, the ball careening down the field, amid fast-moving knees, legs, and elbows. Both teams followed in hot pursuit, when suddenly a panicked shout came from one of the stars on the Danish team. Their own goalie, he had noticed, was paying no attention at all!

Roused by the yell from his brother Harald, the goalie suddenly snapped into action, blocked the ball, and stopped the German drive. Meanwhile, the mathematical problem that he had been figuring on the goalpost waited for another onslaught by the tireless and unstoppable Niels Henrik David Bohr.

The story is one that citizens of Copenhagen love to tell about the rangy, heavy-jowled young physicist with the kind eyes, soft voice, and razor-sharp mind—it's somehow symbolic of Niels Bohr's uncanny ability to be everywhere and do everything at once. And to bring out the best in everyone around him at the same time.

Far from the clichéd image of the absentminded, unworldly scientist (as the public tended to imagine his friendly antagonist Albert Einstein), Niels Bohr loved people. He also felt strongly about issues and delighted in intense discussion and argument. He was aggressively and good-naturedly engaged in the world. Admittedly, there were times when Bohr appeared absentminded—one famous physicist today is the proud owner of a piece of charred chalk that Bohr tried to light, mistaking it for his cigar while lecturing at a blackboard. But not because he lived in an ivory tower—rather, because he had focused so intensely and completely on communicating some difficult idea to a befuddled but absorbed listener.

Next to science and sports, Niels Bohr loved most to talk. In fact, he needed to talk. Many of his friends and colleagues joked that conversation with Niels Bohr was often one-sided, with Bohr doing most of the conversing. Talking was as important as breathing to Bohr. Many scientists do their best work alone at a blackboard or in a lab. Some develop their most profound ideas lost in their own thoughts while walking through gardens or woods, or even lying in bed in the middle of the night. Niels Bohr did his best thinking in the heat of discussion.

Bohr had learned the technique as a child, sitting with his brother on the floor of the family living room in Copenhagen and listening to his father and his friends talk. It wasn't just ordinary everyday conversation that Niels and his brother Harald listened to. Their father, Christian, was a professor of physiology at the University of Copenhagen, an articulate and famous physician deeply interested in science, art, and philosophy; his friends were among the intellectual elite of Copenhagen.

Christian Bohr was politically and socially progressive. Considered somewhat radical in his time, he was a religious skeptic and an early advocate of women's rights. But he still held a deep reverence for the past and its great thinkers. One of his particular favorites was the German philosopher Goethe, whose writings he often read aloud to his children. He was also a sports enthusiast who kept up with all of the latest scores in English football (soccer) games and was instrumental in introducing soccer, then little known in Denmark, to a wider audience.

The lively Bohr household was kept running smoothly by Niels's mother, Ellen, a gentle woman whose quick intelligence and loving nature embraced not only the immediate family but the many visitors who came to the Bohr home. As a result, Niels and his siblings grew up in an intellectual and social center for many of Copenhagen's liveliest minds.

   Niels's father was also the son of a teacher, Henrik George Christian Bohr—Niels's paternal grandfather—who had headed an elementary school on the island of Bornholm, in the Baltic Sea south of Sweden. Christian was born in Copenhagen in 1855 and published his first scientific paper at the age of twenty-two, taking his medical degree in 1878 and receiving his Ph.D. in physiology two years later. In 1881 he married one of his students, Ellen Adler, the daughter of a prosperous Jewish banker and politician. Their first child, Niels's elder sister, Jenny, was born on March 9, 1883. Their second child, Niels Henrik

The Bohr children—clockwise from left, Jenny, Harald (on his mother's lap), Ellen Adler Bohr, and Niels at the age of four. (Niels Bohr Archive, Copenhagen)

David Bohr, followed soon after, on October 7, 1885. The third, Niels's beloved younger brother, Harald, was born on April 22, 1887.

Christian and Ellen spent portions of the early years of their marriage in the large and stately town house owned by the Adler family at 14 Ved Stranden in the heart of Copenhagen. Here Jenny and Niels were born. It was a charming place, surrounded by tiny shops and bookstores and colorful, cluttered rows of red-tile-topped houses. Here young Niels first began to form his strong and loving attachment to Denmark.

Across the street from the Bohr home stood a statue of Bishop Absalon, the Danish soldier, statesman, and prelate who had originally founded the town of Copenhagen as a fortress in 1166. Christianborg Palace, the seat of the government, faced the family's impressive stone-fronted house, and only a short distance away the mooring known as Christianborg Slot was home to the colorful fishing boats that traveled the busy Frederiksholms Canal to the harbor. Despite its damp and often rainy weather, Copenhagen was a bustling and picturesque city. It captured Niels early and held him spellbound throughout his life.

The Bohr family was well-to-do. In 1886, shortly after Christian was appointed associate professor of physiology at the University of Copenhagen, the young family moved into a luxurious and spacious apartment adjoining the university, previously occupied by Christian's predecessor. It was here, at Bregade 62, that Harald was born, and where Niels and his brother and sister lived happily through the remainder of their childhood days.

Christian Bohr encouraged his children by providing opportunities for them to explore their interests. Their summers were spent running and playing at their grandparents' summer home, Naerumgaard, just north of Copenhagen. Niels and Harald often passed their after-school hours with woodworking projects in a shop set up for them by their father in their home. The Bohr children grew up in an environment that encouraged independent development, humane compassion, and culture. Above all, Christian and Ellen Adler Bohr passed on a great love of knowledge and its pursuit.

At a young age, Niels seemed to have a good conceptual sense of the physical world, with a remarkable early interest in reflecting it accurately. One day when he was about three years old, Niels was walking with his father, when the physiology professor began exclaiming about the way a nearby tree's trunk extended up into branches and the branches into twigs, hung with leaves. "Yes,"said little Niels, "but if it weren't like that, there wouldn't be any tree!" And as a fifth-grade student, he painstakingly completed an assigned drawing of a house, only after counting the number of pickets in the fence and representing them exactly in his carefully drafted production. While many people considered Harald to be the brighter of the two boys, Christian Bohr recognized Niels's unique qualities of imagination, always maintaining that Niels was "the special one of the family."

Little, if any, jealousy existed between the two brothers. They could never say enough in praise of each other, and were inseparable in childhood and best of friends throughout adulthood.

In later life, Niels's best friend, Ole Chievitz, would assert, when asked what characteristics of Niels Bohr he would rate the highest: "His goodness . . . Let us not give examples. Bohr would not care for that. You must be satisfied with my word when I tell you that he is as good in big things as in small. I am not exaggerating just because it is his birthday when I say that I consider him the best human being in the world."

But as a child, and even as a young man, he certainly was not a saint. He got into fights and angrily knocked flat anyone who offended his keen sense of justice—often administering a black eye or two. As a boy, young Niels was perpetually in motion, described by those who knew him as a "tornado," full of energy and mischief.

Niels was not always first in his class, according to the memories of his classmates, but he was regularly third or fourth from the top. Although he was outstanding in mathematics and science, for him, writing assignments were another matter, and writing remained an agonizing task throughout his life. Most difficult of all was concluding a piece. For Bohr, there was always more to be said, and the convention of ending with a summary baffled him completely. To him, summaries seemed redundant, repeating unnecessarily, he thought, what had already been said. He finished one early childhood essay on the topic of metals with the sentence "In conclusion, I would like to mention uranium." In later life, someone once asked his brother Harald, by then a well-respected mathematician, how it happened that he had succeeded in building a career as a lecturer, whereas Niels

never had. Harald replied, "Simply because at each place in my
lecture I speak only about those things which I have explained before,
but Niels usually talks about things he means to explain later."

All three Bohr children continued their education at the university.
Jenny studied history at the University of Copenhagen and English at
Oxford. After passing a teachers' examination in 1916, she pursued a
career teaching history and Danish. For Harald and Niels, too, a
university education was the natural, expected follow-up to comple-
tion of their studies at Gammelholm School, the elementary and high
school they both attended.

In 1903 Niels entered the University of Copenhagen, where he and
his brother played on the soccer team. Harald was definitely the better
soccer player—later playing halfback on the 1908 Danish Olympic
team (winner of the silver medal). Niels, though an avid and excellent
athlete, only served as reserve goalkeeper on the university team. But,
for Denmark, the Bohr brothers were champion "footballers."

During these years, Niels developed a natural ear for poetry and
memorized many stanzas in German and Danish. In addition, he
enjoyed reading philosophy—including the Danish philosopher
Søren Kierkegaard—and fiction, and he especially valued a little book
by the Danish writer Poul Martin Møller, *Tale of a Danish Student*.
Bohr was always fascinated with the attempts Møller's student had
made to sort out the many dualities inherent in life. As Møller wrote,

> Thus on many occasions man divides himself into two persons, one of
> whom tries to fool the other, while a third one, who in fact is the same
> as the other two, is filled with wonder at this confusion. In short,
> thinking becomes dramatic and quietly acts the most complicated plots
> with itself, and the spectator again and again becomes actor.

The last line, in particular, often found its way into Bohr's conversations. And this fascination with duality—two things at once—would
fill his thoughts, even his thoughts of physics, throughout his life.

At the university, Harald and Niels took a philosophy class from
their father's friend Harald Høffding and, following their father's
tradition, formed a discussion group with some of their classmates.
They met over coffee or beer and often discussed the problems raised
in Høffding's classes—their conversations frequently dominated by

Harald (left) and Niels Bohr as students. (Niels Bohr Archive, Copenhagen)

Niels, who held forth in his often-mumbling manner, his listeners constantly entreating him with pleas of "Louder, Niels."

By 1905, at the age of 19, Niels was ready to plunge into the world of physics. The Royal Danish Academy of Sciences and Letters had proposed an award for the best paper on the surface tension of liquids, and Niels decided to enter. Lord Rayleigh, who was a major figure in physics at the time, had proposed that it was possible to figure out the surface tension of a liquid if a few factors were known. If one could take measurements of the length of the waves that formed on a jet of the liquid—such as might be produced by the nozzle of a hose—and one knew the speed of the jet and its cross section, then it would be possible to determine the amount of tension present on the surface.

Niels's father, delighted with his son's ambition, gladly volunteered the facilities of his physiology laboratory for the project. For months

Niels spent long night hours—when no one would disturb the quiet he needed—producing perfect streams of water. By heating and drawing his own glass tubing, he had devised a method for producing a jet of water that would always have the same speed and cross section. And he spent hours measuring and remeasuring.

Finally the deadline was nearing, and he had not begun to write. His father—seeing that Niels had become mesmerized by his experiments and probably would never finish without intervention—packed his son off to the Adler summer home to write, far away from interruptions and temptations to experiment further. The ploy worked. His triumphant paper, though inconclusive, raised important questions, extending Lord Rayleigh's basic theory about the surface tension of liquids. Niels Bohr was declared one of two winners in the competition, and four years later, in 1909, the British Royal Society published the paper, translated and in modified form, in its *Philosophical Transactions,* a heady triumph for such a young scientist, especially for work done while still a college undergraduate.

It was an auspicious start. Bohr had begun, by accident, with a subject that would lead unexpectedly, 35 years later, to an understanding of how an atomic nucleus might split, opening a door to the tremendous explosive power of an atomic bomb and the development of nuclear energy.

By this time Bohr had begun to immerse himself in the key issues of modern physics. The hottest new topic in the journals, he found, was radioactivity—the talk of all the scientific circles in France, Germany, and England. And so for his required student presentation on some aspect of his field, Bohr settled on radioactivity.

Much of the hubbub had begun when Niels was about 10. During the 1890s, scientists were fascinated with cathode rays. If nearly all the air from a glass tube sealed with metal plates at each end is removed and then the metal plates are connected to a battery, the emptiness in the glass tube glows. The glow stretches from the negative (cathode) plate to the positive (anode) plate, and that's why the rays so produced are called cathode rays. (With a few adjustments, this is the principle of today's television tubes.)

In 1895 in Bavaria, Wilhelm Conrad Röntgen was preparing for some experiments with cathode rays when he discovered a strange new ray being emitted through the heavy black paper he had placed over his cathode ray tube. The ray caused a piece of paper coated with barium platinocyanide (used in photography) to glow. He found that the new ray could travel through walls. When he placed his hand in the ray's path, it passed through the flesh of his hand but not the bones, which appeared as dark rods in the brightness. He called the ray "X ray," "x" for unknown.

In England, at the Cavendish Laboratory at Cambridge University, the physicist J. J. Thomson began investigating these issues. By 1897 Thomson had proved that cathode rays were not light waves, as many scientists had assumed. These were negatively charged particles, he showed, that virtually boiled off the negative cathode and sped straight for the positive anode, propelled by electrical attraction. He showed that a narrow beam of cathode rays could be deflected in both an electric and a magnetic field. What, then, could the cathode ray be composed of?

Thomson came up with an idea that shook the very foundation of physics and chemistry. These rays, he asserted, must be made up of particles that he called "negative corpuscles" (later renamed "electrons" by the Dutch scientist Hendrik Lorentz). What's more, Thomson said, they must be part of matter itself, part of the atom.

To physicists and chemists alike, all over the world, these pieces of evidence began the call for a new vision of the atom, the indestructible building block of all matter. The atom was a very old idea, dating back to the fifth century B.C., when two Greek thinkers, Leucippus and his student, Democritus, first described it. All matter, they said, was made of fundamental particles that could not be split into parts. They called these particles "atoms." (The Greek word *atomos* means "indestructible.") These tiny, indestructible particles were the very essence of matter, they said. And each element, such as gold, silver, hydrogen, and so on, represented a single substance composed entirely of only one type of atom. Over the centuries this atomic theory seemed to fit experimental results consistently, and by the early 19th century most scientists had concluded that it depicted an accurate view of reality. By 1895 about 76 of these elements were known, and a Russian scientist by the name of Dmitry Mendeleyev had observed common traits among some of them when he arranged them by atomic weight (the weight of a single atom of an element). He called the arrangement he came up with the "Periodic Table of Elements." With only a few alterations, his periodic table is still in use today.

But now J. J. Thomson measured his electrons and found that they were much lighter than hydrogen atoms, the very smallest atom of all. (His first measurement put the new particle at 1/770th the mass of hydrogen, but a later measurement of 1/1840th proved more accurate.) However, according to the prevailing atomic theory, nothing could be smaller than the hydrogen atom. He also found that it didn't matter whether the cathode tube contained a vacuum or was pumped with any kind of gas you could imagine. His electrons were always the same.

Meanwhile, Röntgen's discovery intrigued Henri Becquerel, a Parisian expert on fluorescence, which is the natural tendency of some substances to glow. Becquerel knew that sunlight can induce fluorescence in some substances, and he decided to try some experiments with various materials. He decided to begin with a uranium salt, but met with a series of cloudy days and so, to save the materials for a sunny day, he wrapped the uranium subtance in paper and stuck the package in a drawer with the unexposed photographic plates. When he retrieved the materials, he was amazed to find that the uranium salt had exposed the photographic plates, even though the salt was completely hidden from any light. He had discovered, quite by accident, that some substances give off another ray, similar to X rays, but not the same. The year was 1898. Marie Curie, a talented young physicist working in Paris at the time, gave the name "radioactivity" to this newfound emanation.

The implications of these discoveries, when closely examined, set late 19th-century science into a great stir. Where did these rays come from? What were they? Thomson had already assailed the idea that atoms were the smallest unit of matter. Now, with the discovery of radioactivity, the Greek atom was called even more into question. Where could these rays be coming from other than from the atoms themselves? If pure samples of an element could emit rays, the atoms must be disintegrating in some way; that is, the rays must be composed of *part* of the atom. But atoms, by the old Greek definition, had no parts. An atom, according to Leucippus and Democritus, was unsplittable. With the discoveries of Thomson, Röntgen and Becquerel, the ancient Greek view of the atom as solid and unsplittable had come to an end.

This was the kind of dilemma physics was facing when Niels Bohr prepared his paper on radioactivity. He had picked out the very area of physics that was at that moment just ready to burst into hundreds of new avenues of investigation.

Niels received his bachelor's degree from the university in 1907 and continued on as a graduate student to work on his master's degree. For his thesis work, he chose the subject of the electron theory of metals. By 1904 Thomson had set forth, based on his discovery of the electron, a revolutionary new vision of the atom—often referred to as the "plum pudding" or "raisin-in-pound-cake" model. An atom, he said, was primarily a positively charged sphere, embedded with negatively charged electrons, much as raisins are embedded in a plum pudding or pound cake. These electrons, researchers began to realize, determined how elements combined, and it began to become apparent that all chemistry was ultimately a matter of electrical attractions. Niels, working very near the cutting edge of his field, was exploring how the theory of electrons related to the characteristics of metals.

Meanwhile, Harald, back from the Olympics, moved steadily forward to work on his own master's thesis in mathematics, finishing before his brother, in 1909. Harald then set off for the university at Göttingen, where the most advanced work in his field was taking place. Niels, who was still laboring over his thesis in June 1909, was both a little jealous and very proud.

Once again, for Niels, writing the thesis would require getting away from the temptation to continue researching and experimenting. Upon his grandmother's death, her home at Naerumgaard had been given away by the Adler family to become a children's home, so it was no longer available as a retreat. Niels went instead to a vicarage located on the island of Fyn, one of the main islands of Denmark. It was the home of the father of Christian Bohr's assistant, Holgar Mollgaard, and it turned out to offer the isolation and opportunity for focus that Niels needed for his writing. There he pored over the books and articles of Thomson, as well as publications by Paul Drude and Lorentz. He succeeded in homing in on assumptions and uncovered a few discrepancies in Lorentz's work (as he wrote parenthetically in a letter to Harald, "you know I have the bad habit of thinking I can find mistakes

Margrethe Nørlund and Niels Bohr at the time of their engagement in 1911. (Niels Bohr Archive, Copenhagen)

in others"). Bohr's innate ability to question and put his finger on fundamental assumptions had already become an important part of his scientist's bag of tools.

Just after submitting his master's thesis, Niels took a holiday to visit a friend, Niels Erik Nørlund, a member of the discussion group that he and Harald had formed. There he met his friend's sister, Margrethe Nørlund, an intelligent, warm young woman with a serene strength and a more than passing interest in science. Niels had met his life partner, and he knew it at once.

As a lifelong friend, Richard Courant, wrote many decades later, after Niels's death, "It was not luck, rather deep insight, which led him to find in young years his wife, who . . . had such a decisive role in making his whole scientific and personal activity possible and harmonious."

Margrethe would become his partner in every way. Their happiness together was always apparent to those who knew them, and their marriage, which would last over 50 years, was unique. As one biographer put it, "Not only by the strength of her great personality and by her knowledge and ability in so many differing fields, but especially by her devotion, Margrethe Bohr became the perfect and indispensable support to her husband."

With the beginning of the school year in the fall of 1909, Harald returned from Göttingen to continue his Ph.D. studies, and the two brothers often studied together. Niels also began work on his doctorate, choosing to continue his work on the electron theory of metals for his thesis.

By January 1910, with Niels still lagging behind, Harald was ready to defend his thesis in mathematics. As is still the case at many universities throughout the world, a candidate for a Ph.D. in Denmark at that time had to answer questions before a faculty committee in defense of the thesis. According to tradition, friends, relatives, and colleagues frequently attended the proceedings in support, and in Harald's case the entire room at the university was packed with a boisterous cheering section composed of his teammates and supporters of his soccer team. If the members of his committee were unsettled by the unusual audience, they showed no sign of it, and Harald passed his candidacy without a hitch, to the joy of the crowd.

For Niels, though, defense of his thesis, on the electron theory of metals, was still more than a year off. Long hours of research and, later, hours of writing once again during a six-week retreat at the vicarage on Fyn lay between him and the moment of victory. In June

1910 he wrote to Harald, "have succeeded only in writing circa fourteen more or less divergent rough drafts." He finally finished that summer, and in August he became engaged to Margrethe Nørlund. But his approaching success—the formal defense of his dissertation the following year—would be marred by the sudden loss of his father.

Christian Bohr had been working late at his laboratory the night of February 2, 1911, after a supper in his home attended by Margrethe Nørlund. He returned home about midnight, when he began to develop sharp chest pains. Recognizing the warning signs of a coronary attack, he called his assistant and family, but the pains subsided. Relieved, he announced that he guessed he would have to give up smoking for a while, but the reprieve was all too short. Moments later he collapsed, a victim of heart failure, just days before his 56th birthday. His remains were cremated and buried in one of Copenhagen's oldest cemeteries near the graves of the physicist Hans Ørsted and the fabulist Hans Christian Andersen.

With the death of Christian Bohr, Niels lost a strong ally, just as he embarked on a career that would have made his philosopher-physiologist father very proud. On May 13, when Niels defended his thesis, the opening page read: "Dedicated with deepest gratitude to the memory of my father."

Niels's introductory remarks at his defense were written in Margrethe Nørlund's handwriting, as most of the rest of his papers would be until he began to find young physicists he could talk into doing the task. Bohr was never a man who thought with pen in hand. He did his best thinking in conversation, pacing about the room, pausing in the middle of sentences and thinking out loud. And, fortunately for science, he was always able to coax willing collaborators into helping him get the words on paper.

On this occasion, however, he was unusually silent.

"Dr. Bohr, a pale and modest young man," reported a newspaper account of the event, "did not take much part in the proceedings, the short duration of which is a record. . . . The words Bohr had written and the questions he had raised were literally so new and unusual that no one was equipped to question them."

That summer, Niels Bohr spent many idyllic hours walking, hiking, and talking with Margrethe and Harald. But as the Nordic days began to shorten in September, he set off for his postgraduate work abroad.

Clearly, he would continue his work where the action was: in Cambridge, England, academic home of J. J. Thomson. As he set sail, he may well have been thinking of lines he often quoted from one of his favorite poems by Hans Christian Andersen:

In Denmark I was born, there is my home,
there are my roots, from there my world unfolds . . .

For Niels Bohr was a young man imbued in Danish culture, but who, from the days when he sat listening to learned discussions at his father's knee, reached always outward to embrace the world beyond.

CHAPTER 1 NOTES

p. 5   "Yes, but if it weren't like that . . ." Quoted by S. Rozental in "Childhood and Youth," in *Niels Bohr: His Life and Work as Seen by His Friends and Colleagues,* S. Rozental, ed., p. 17.

p. 5   Niels is "the special one." Quoted in Rozental, p. 15.

p. 5   "His goodness . . ." Ole Chievitz, in a press interview on the occasion of Bohr's 60th birthday, quoted in Rozental, ed., pp. 21–22.

p. 5   "In conclusion . . ." Quoted in Ruth Moore, *Niels Bohr: The Man, His Science, and the World They Changed,* p. 12.

p. 6   "Simply because . . ." Harald Bohr, quoted by R. Courant, in "Fifty Years of Friendship," in Rozental, ed., *Niels Bohr,* p. 303.

p. 6   "Thus on many occasions . . ." Møller, quoted by Richard Rhodes, *The Making of the Atomic Bomb,* p. 59.

p. 7   "Louder, Niels." Vilhelm Slomann, a member of the group who wrote an article in honor of Bohr's 70th birthday, quoted by Rozental, ed., p. 25.

p. 11   "you know I have the bad habit . . ." Quoted in Ruth Moore, p. 27.

p. 13   "It was not luck, rather deep insight . . ." Courant, in Rozental, ed., p. 304.

p. 13   "Not only by the strength . . ." Rozental, ed., p. 37.

p. 14   "have succeeded only . . ." Bohr, July 26, 1910, *Collected Works,* vol. 1, p. 51.

p. 14   "Dedicated with deepest . . ." Bohr, *Collected Works,* vol. 1, p. 295.

p. 14   "Dr. Bohr, a pale and modest . . ." Bohr, *Collected Works,* vol. 1, pp. 98–99.

p. 15   "In Denmark I was born, . . ." Hans Christian Andersen, translated by Abraham Pais, *Niels Bohr's Times: In Physics, Philosophy, and Polity,* p. 39.

2

ENGLAND, RUTHERFORD, AND THE MYSTERIES OF THE ATOM 1911–1912

When Bohr arrived in Cambridge in late September 1911, he felt exhilarated. He walked the very paths Isaac Newton and James Clerk Maxwell—two towering genuises of physics—once had trod, and the days that lay ahead seemed to hold boundless opportunity.

Prior to his death, his father had helped arrange a grant for the young physicist's postgraduate work, to be pursued under the guidance of none other than Joseph John Thomson—"J. J." to the world (his students included)—discoverer of the electron and head of the prestigious Cavendish Laboratory. Niels lost no time making contact. As soon as he had found a place to live and done the minimal unpacking, he set off to see J. J., with a translation of his own Ph.D. dissertation into English (done, rather badly, by a friend) and a copy of one of J. J.'s recent papers under his arm.

J. J., who was 56 at the time, had already been head of the Cavendish for 27 years—having taken the place of Lord Rayleigh at the age of 28. He was only the third head of the prestigious laboratory—the first director had been James Clerk Maxwell. Thomson was an intense-looking man who wore wire-rimmed glasses and a slightly drooping mustache. His hair, a bit long, swept across the crown, and his proper tweed coat and winged collar tended to look disheveled, the requisite tie and cuff links in place but, like his cluttered desk, not necessarily in order. But, for all his erudition and fame—he had won the Nobel Prize only five years earlier for his work on the electron and was knighted in 1908—his manner was genuinely warm and sociable. As

Niels Bohr made his way along the darkly paneled halls of the Cavendish Laboratory and entered Thomson's office, he might have been overcome with a sense of awe, had it not been for his new mentor's disarmingly friendly manner.

Young Bohr's English was halting, but he began by opening Thomson's paper and pointing out a few places where he thought Thomson had possibly gone wrong. Bohr finished by offering Thomson his dissertation on the application of electron theory to metals. He hoped that Thomson might read the thesis and discuss it with him. Perhaps he even might be willing to help Bohr get it published in England. Thomson accepted the dissertation cordially and placed it on top of a stack of papers on his desk.

Bohr went back to his rooms elated. That evening he wrote to his brother:

> 29 Sept. 1911
> Eltisley Avenue 10,
> Newnham, Cambridge
>
> Oh Harald!
> Things are going so well for me. I have just been talking to J.J. Thomson and have explained to him, as well as I could, my ideas about radiation, magnetism, etc. If you only knew what it meant to me to talk to such a man. He was extremely nice to me, and we talked about so much; and I do believe that he thought there was some sense in what I said. He is now going to read [my dissertation] and he invited me to have dinner with him Sunday at Trinity College; then he will talk with me about it. You can imagine that I am happy. . . . I now have my own little flat. It is at the edge of town and is very nice in all respects. I have two rooms and eat all alone in my own room. It is very nice here; now, as I am sitting and writing to you, it blazes and rumbles in my own little fireplace.

However, time passed, and Bohr heard no word from Thomson. Had he perhaps gone too far, calling the great physicist's theory into question? He wrote to Margrethe, "I wonder what he will say to my disagreement with his ideas." A few weeks later he wrote her again: "I'm longing to hear what Thomson will say. He's a great man. I hope he will not get angry with my silly talk."

During his leisure hours in Cambridge, Niels called on former students of his father's, attended teas and gatherings sponsored by the ladies of the English academic community, read the works of Charles

Dickens with dictionary in hand to improve his English (remembering the Dickens stories his father used to read when he was a boy), and enjoyed long walks through the countryside.

After one of these autumn excursions, he wrote enthusiastically to Margrethe, ". . . and then I went on the loveliest walk for an hour before dinner across most beautiful meadows along the river, with the hedges flecked with red berries and with isolated wind- blown willow-trees—just imagine all this under the most magnificent autumn sky with scurrying clouds and blustering wind . . ."

Always athletic, Bohr also joined a soccer club and, as the weather got colder, enjoyed ice skating. Harald came for a visit at Christmas.

But as far as physics went, Bohr spent his time working on a lusterless project on cathode ray production, suggested by Thomson. It was not going well and seemed to hold little promise of yielding fruitful results. Try as he might to sound upbeat in his letters, he became restless. No one he knew in England understood Danish, and Bohr's usual roundabout way of speaking and his soft voice didn't help his less-than-complete command of English. His colleagues couldn't understand, for example, what he meant by "loaded" electrons (he meant "charged"), and his miscommunications were often comical. On other occasions the handicap must have been very frustrating. In one instance, at a meeting, Thomson waved aside comments by Bohr as useless, only to restate the same ideas in different words.

Finally Bohr could stand the suspense no longer, and he went again to visit Thomson. The interview was again cordial, but the dissertation remained unread, still in a stack on the director's desk. Unknown to Bohr, Thomson was notorious for neglecting student papers and correspondence. Also, Thomson never said so, but his lack of interest may have grown out of the fact that since completing his own work on electron theory, he had turned to other subjects and was no longer as involved as he once was. The fact is, he never did read Niels Bohr's dissertation, which would not be published in a good English translation until 1972.

Bohr left the director's office deeply discouraged. Things were not going so well at Cambridge after all. Years later he would comment, "The whole thing was very interesting in Cambridge but it was absolutely useless."

Ernest Rutherford in 1902, nine years before meeting Niels Bohr. (AIP Emilio Segré Visual Archives)

But Bohr was not about to let his one year in England go so badly. He resolved to make a change, and he looked toward Manchester, which may have seemed an unlikely choice—compared to Cambridge, the University of Manchester was like meat and potatoes next to caviar and wine. But for a student of physics, Manchester had one great asset: Ernest Rutherford.

Rutherford was an exuberant man with big hands, a big walrus mustache, and a booming voice. When Ernest Rutherford had arrived on scholarship at the Cavendish Laboratory from his native New

Zealand, he was a brash 24-year-old with strong opinions, plenty of ambition, and no money. He also had the qualities of a fine physics experimentalist, which Thomson recognized immediately.

Rutherford couldn't have arrived at a better time. Three months later Röntgen discovered the X ray; three months after that Becquerel discovered radioactivity. And a year later Thomson discovered the electron. By this time Rutherford was working in Thomson's lab on questions surrounding radioactivity.

In 1898, when Marie Curie reported her discovery of new radioactive substances, one of them, radium, proved to be several million times more radioactive than Becquerel's uranium salts. Physicists and chemists became intensely interested in the subject, and Rutherford was in on the ground floor. In his work with J. J., he had learned just the right techniques for discovering more about radioactivity, and he began studying rays emitted by uranium. What he was trying to do was dissect the atom. In the process he found out something big: Uranium gave off two very distinct types of radiation. He called one "alpha" radiation (after the first letter of the Greek alphabet). Alpha rays were easily absorbed by matter. The other he called "beta" radiation (after the second letter of the Greek alphabet). This was a far more penetrating ray. (Paul U. Villard from France discovered a third even more penetrating ray in 1900, which was later named "gamma" by Rutherford, after the third letter of the Greek alphabet.)

The competition was keen, as Rutherford wrote home to his mother: "Among so many scientific bugs knocking about, one has a little difficulty in rising to the front." But Rutherford was never one to let anything, much less a little competition, stop him. In later years his students nicknamed him "the Crocodile" because, as one explained, "the crocodile cannot turn its head . . . it must always go forward with all devouring jaws." He was tireless in every quest and loved his role as one who put endless questions to nature. His great success, as Bohr would later say, came from his knack of forming questions in such a way that they could produce the most useful answers.

A consummate experimentalist, Rutherford generally had little use for theorists. "They play games with their symbols," he said, "but we [experimentalists] must turn out the real facts of Nature." He had a particular talent for experimental design and an uncanny ability to pick out one significant fact from a mass of confusing detail. As one colleague remarked, "With one movement from afar, Rutherford so to speak threaded the needle the first time."

Rutherford completed this early work on alpha and beta radiation at the Cavendish, but by the time it was published in 1899, he had accepted an appointment as professor of physics at McGill University in Canada. There, in 1900, he noticed that thorium gave off a radioactive gas, and he enlisted the help of a young chemistry professor, Frederick Soddy, to help him identify the gas. When Soddy examined it, he found that the gas had no chemical properties at all. Only one element fit that description, the chemically inert gas argon. From this evidence, according to Soddy, they reached "the tremendous and inevitable conclusion that the element thorium was slowly and spontaneously transmuting itself into argon gas!" Rutherford and Soddy had discovered the stunning fact that, through radiation, radioactive elements change themselves into other elements.

In the following years, Rutherford continued to unravel the threads of the complex mysteries surrounding the transmutation of the radioactive elements. This work done at McGill University earned him, in 1908, a Nobel Prize, oddly enough, not in physics but in chemistry.

By 1908 Rutherford had returned to England, having accepted the position as professor of experimental physics at the University of Manchester that he held four years later, when Bohr arrived. At Manchester, his students tended to be mature, and most of them already held degrees. One of them, a young German physicist named

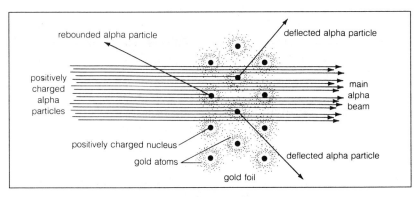

Figure 1. Rutherford's discovery of the nucleus. (From Jacqueline D. Spears and Dean Zollman, *The Fascination of Physics.* Benjamin/Cummings, 1985.)

Hans Geiger, teamed up with Rutherford, and together they began bombarding thin pieces of gold foil with alpha particles. Most of the alpha bombarders passed right through the foil, which was exactly what the experimenters expected, based on Thomson's "plum pudding" model of the atom. But some of the alpha particles struck the gold foil and were deflected at a sharp angle—often 90 degrees or more! This amazed Rutherford, who remarked, "It was almost as incredible as if you fired a 15-inch shell at a piece of tissue paper and it came back and hit you." The Thomson model of the atom clearly required rethinking.

Early in 1911 Rutherford came buoyantly to Geiger. "I know what the atom looks like!" he exclaimed. Based on his results, Rutherford put together a new idea of the atom: What if all the positively charged particles in the atom were not spread like a fluid throughout the atom, but lumped together in the center in one tiny core area? This "nucleus" (a term he came up with in 1912) was thousands of times heavier than an electron, so most of the atom's mass would be contained there, composed of positively charged particles. An equal number of negatively charged electrons would be found in motion somewhere outside the nucleus. The rest of the atom would be empty space. If the nucleus was the size of a marble, sitting in the center of the field in an empty football stadium, its electrons would be perched on the outer walls of the stadium. It was a compelling idea—a sort of tiny planetary system that mirrored the larger Solar System we live in. In the bombardment experiments, most of the bombarding alpha particles would pass straight through the comparatively vast areas of empty space. But a few would hit the nucleus of positively charged particles, be repelled (since alpha particles are also positive) and bounce back.

Rutherford's discovery of the atomic nucleus was stunning, and it would ultimately earn him the title of "the Newton of atomic physics." But not many people paid much attention to it at the time. In fact, in the fall of 1911, Rutherford attended an international conference of the world's leading physicists, the first of a series of prestigious conferences known as the Solvay conferences (after the man who organized the first one). And at that conference he made no mention of his new atomic model.

When Bohr arrived in Cambridge, however, he already knew about Rutherford's work with radioactivity from the journals he had read

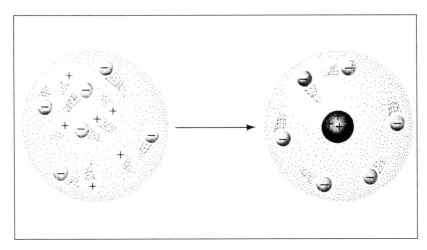

Figure 2. Early models of the atom: Based on his discovery of the electron, J. J. Thomson suggested in 1898 (left) that atoms were spheres of positively charged matter with negatively charged electrons embedded in them—something like raisins in pound cake. In 1911 Ernest Rutherford came up with the idea that each atom consisted of a tiny positive nucleus with electrons circling somewhere outside it. (From G. Tyler Miller Jr., *Chemistry: A Basic Introduction,* 2nd ed. Belmont, Calif.: Wadsworth Publishing Company, 1981.)

and about his emerging new model of the atom, which Bohr considered to be at the cutting edge of physics. (Of course, not everyone, least of all Thomson, agreed.) Now Bohr thought that perhaps it was time for a visit to Manchester. He arranged to see an old friend of his father's there who was also acquainted with Ernest Rutherford. In this way Bohr got to meet Rutherford, and the two hit it off immediately.

Bohr saw Rutherford again a few weeks later at the December 8, 1911 Cavendish Research Students' Annual Dinner. As one of J. J.'s "old students," Rutherford was invited to give a speech. The mood was jolly and the menu was posh—including turbot, shrimp, plover, mutton, turkey, pheasant, and plum pudding—no doubt with raisins. The program included rollicking limericks sung with more gusto than art, including "My name is J. J. Thomson and my lab's in Free School Lane/There's no professor like J. J. my students all maintain" and "For an alpha ray/Is a thing to pay/And a Nobel Prize/One cannot despise/And Rutherford/Has greatly scored/As all the world now recognize."

Rutherford spoke enthusiastically and excitedly, and Bohr was impressed with the charm, power, and enthusiasm with which he described the work of C. T. R. Wilson, an experimentalist who had invented an instrument known as a cloud chamber, used for tracking atomic particles. With his cloud chamber, Wilson could see the paths of charged particles visible as lines of water droplets hovering in a supersaturated fog. He had just photographed the evidence of alpha particles in his cloud chamber as they scattered from interactions with nuclei, showing scatterings very similar to those that had preceded Rutherford's discovery of the nucleus just a few months before.

Sometime shortly thereafter, having consulted with Harald during his Christmas visit, Bohr made arrangements with Rutherford to transfer to Manchester to finish his year. Rutherford welcomed the idea, although he advised the eager young scientist to finish out his work with Thomson first. This Bohr did, explaining to Thomson that he "should be glad to know something about radioactivity," and then he set off for Manchester in March 1912.

Niels Bohr was perhaps the only theorist ever to hit it off with Rutherford. "Bohr's different" was the explanation Rutherford offered in his usual brusque way. "He's a football player." In a way that quick summary represented something important about Bohr as a physicist: He was in every way a man who lived in the real world. He prided himself on the literal-mindedness that helped him maintain his connection with the physical world while he thought about the ideas of physics.

But Bohr and Rutherford were a strange pair, Rutherford brash and exuberant, Bohr speaking not much above a whisper, digging in his mind for the perfect word and, in the words of fellow physicist C. P. Snow, "on not finding it, . . . pauses, minutes long, in which he reiterated a word which was clinging to his mind." The contrasts between these two men and their problem-solving methods epitomized the differing styles of experimental and theoretical physicists that emerged fully in the first decades of the century—no two people better typified the dichotomy than Bohr and Rutherford. Bohr the ruminator, with enormous powers of concentration, thought things through as he talked and frequently would hit upon an idea spontaneously in the middle of a conversation. He had no mechanical talent, however, for coaxing resistant laboratory apparatus into smooth operation and exhaustive experimentation. Rutherford, by contrast, had the intractable persistence needed to pursue a course of action and see it through to its outcome, but lacked Bohr's ability to day-dream with purpose. In solving the "serious problems of physics," the physicist Otto Frisch would later recall that Bohr "moved with the skill of a spider in apparently empty space, judging accurately how much weight each slender thread of argument could bear."

For Bohr, Manchester was a wonderful contrast to Cambridge. Granted, it was not such a pretty place. The bustling industrial city was noisy, its buildings blackened and sooty from the belching factory smokestacks. Horse-drawn drays crowded the busy, cobblestone streets. But the physics laboratory at the university was rapidly becoming one of the most productive in the world. And, at its head, Rutherford created around himself an atmosphere of intellectual excitement and openness in which young Bohr flourished—and found the seeds for his own teaching style in later years. One of Rutherford's collaborators, E. Andrade, once wrote this description of his style: "Although there was no doubt as to who was the boss, everybody said what he liked without constraint . . . He was always full of fire and infectious enthusiasm when describing work into which he had put his heart and always generous in his acknow-ledgement of the work of others."

In Manchester, a young Hungarian Bohr's age, Georg von Hevesy, took Bohr under his wing and introduced him to the life of the lab. Hevesy had followed Rutherford to Manchester from McGill the year before and was working on a project—proposed by Rutherford—to separate radioactive decay products from their parent substances. It turned out to be an extremely difficult challenge, one that gave Hevesy another idea. Over the next several decades he developed the science of using radioactive tracers in medical and biological research. Hevesy became a good friend of Bohr's, and he also had an extensive knowl-edge of chemistry—especially radiochemistry—which turned out to be exactly what Bohr needed for the work ahead.

Bohr began by taking an eight-week laboratory course, with Geiger as one of the instructors, in the experimental methods of radioactive research. The course ended May 3. After that, Rutherford set him to work studying the absorption of alpha particles in aluminum. Bohr wrote to his brother, " . . . Rutherford is a man you can rely on; he comes regularly and enquires how things are going and talks about the smallest details. . . . Rutherford is such an outstanding man and really interested in the work of all the people who are around him. . . ."

The equipment was primitive by modern standards, and the work was tedious, but there was always a sense of discovery in the air. Geiger once described it as "the gloomy cellar," which Rutherford had fitted with the delicate apparatus he used for the study of the alpha rays. "Rutherford loved this room," wrote Geiger. "One went down two steps and then heard from the darkness Rutherford's voice reminding one that a hot-pipe crossed the room at head-level, and to step over two water-pipes. Then finally, in the feeble light one saw the great man himself seated at his apparatus."

Each afternoon in the lab, work was set aside for tea. Rutherford would come in, sit down, and talk. The lab group avidly discussed politics and sports and, of course, work. Ideas always were exchanged freely at these daily get-togethers. So much was happening in physics that no one was afraid that someone else would take his idea and publish it first. There were plenty of vital topics for everyone.

Outside the Manchester lab, though, not many scientists had made much of Rutherford's discovery of the nuclear atom, partly because Rutherford himself had not made a strong case for it.

During this period, Bohr began thinking about some of what he had learned at Manchester. Hevesy mentioned to him that the number of radioactive elements far outnumbered the available spaces on the periodic table. Bohr remembered that, according to Soddy, most radioactive elements were not new elements, just variants of natural elements that were already known. (Soddy later came up with the name "isotopes," the term now used for such variants.) For example, uranium 235, which is radioactive and very rare, is an isotope of uranium 238. Bohr realized that the natural elements should be organized in the periodic table not as they had been, according to the weight of an element (its atomic weight) but according to the number of protons in its nucleus (its atomic number)—and that the atomic number for the radioelements must be the same as the natural elements with which they were chemically identical. From these ideas, he roughed out what is now known as the radioactive displacement law. Basically, this law states that when an element emits an alpha

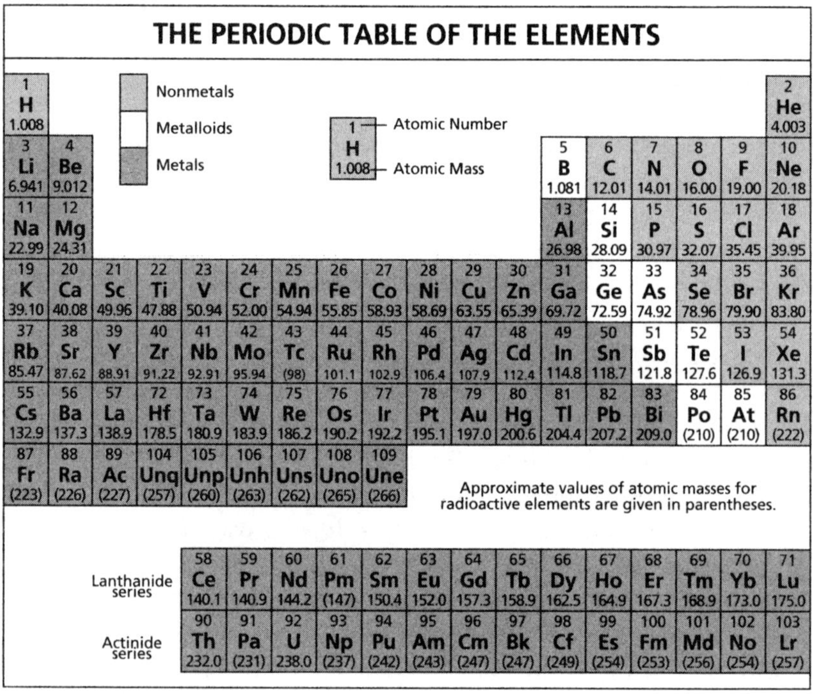

THE PERIODIC TABLE OF THE ELEMENTS

Figure 3.

particle (a helium nucleus, having the atomic number 2) through radioactive decay, it moves two places to the left on the Periodic Table (down in atomic number); when it emits a beta particle, it moves to the right one place (up in atomic number because, as an energetic electron, the beta particle leaves behind an extra positive charge in the nucleus).

Bohr was excited when he had worked out his calculations, and he ran to report to Rutherford. To his surprise, Rutherford was cautious. Bohr later recalled, "Rutherford . . . thought that the meagre evidence [obtained up to that point] about the nuclear atom was not certain enough to draw such consequences. And I said to him that I was sure that it would be the final proof of his atom." As it turned out later, Bohr was right, but because of Rutherford's caution, he didn't publish his ideas.

However, as Bohr was waiting for some radium he needed for another alpha-scattering experiment, he came across a paper on the energy loss that occurs when alpha particles are not scattered by a collision with a nucleus but pass through a metal instead (as the vast majority of them do). It was written by another student of Rutherford's, Charles Galton Darwin, the grandson of the great evolutionist Charles Robert Darwin. Darwin had figured, correctly, that most of the energy loss was caused by encounters the alpha particles made with electrons. But he hadn't considered how or even whether the electrons might be moving. Instead, he assumed that the electrons were free. As Niels wrote to Harald, concerning Darwin's paper:

> It seemed to me that it was not only not quite right mathematically (this was however rather trifling) but quite unsatisfactory in its basic conception . . . I have worked out a theory about it, which, however modest, may perhaps throw some light upon a few things concerning the structure of atoms. . . . I am considering publishing a little paper about it.

Bohr began his paper but didn't finish it until after he had returned to Denmark, partly because he found some other ideas to think about that were more exciting. By mid-June 1912, he had worked out some more calculations and went to see Rutherford again. In a letter to Harald dated June 19, Bohr reported what happened:

> It could be that I've perhaps found out a little bit about the structure of atoms. You must not tell anyone anything about it, otherwise I certainly could not write you this soon. If I'm right, it would not be an indication of the nature of a possibility . . . but perhaps a little piece of reality. . . . You understand that I may yet be wrong, for it hasn't been worked out fully yet (but I don't think so); nor do I believe that Rutherford thinks it's completely wild; he is the right kind of man and would never say that he was convinced of something that was not entirely worked out. You can imagine how anxious I am to finish quickly.

Bohr had found himself drawn to questions surrounding the Rutherford model of the atom. There was something wrong with this atomic model based on the discovery of the nucleus, the model that depicted the atom as a miniature Solar System.

As Rutherford had pointed out, the electron was attracted to the opposite electrical charge of the nucleus—the electron having a negative charge, the nucleus a positive charge. This explained why, as the positively charged alpha particles sped toward the nucleus, they had veered away so dramatically—because like charges repel.

In this configuration, an electron would move in an elliptical orbit around the nucleus, in the same way that the planets move around the Sun. But the idea of a moving electron seemed impossible—impossible because, according to the laws of electricity, a moving charge must produce electromagnetic radiation (light, for example). If all electrons produced radiation, then all matter would radiate, which it doesn't. An even bigger problem lay in the fact that as radiation is released by moving electrons, they should lose energy and rapidly spiral into the nucleus. The principle is similar to the way a satellite losing energy from the drag of the Earth's atmosphere will eventually crash into the Earth. But Rutherford's electrons would crash into the nucleus in as little as 10^{-10} seconds, or one 10-billionth of a second. Atoms, according to Rutherford's model, in short, must be very unstable. But, of course, atoms are not unstable. This was the dilemma that Bohr undertook to solve, and he believed he had an answer.

He went back to his rooms to continue his calculations, but he was running out of time. His departure from Manchester was planned for the end of July and he was eagerly awaited in Denmark, where he was planning to marry Margrethe Nørlund on August 1. On July 17 he wrote Harald again:

> I believe I have found out a few things; but it is certainly taking more time to work them out than I was foolish enough to believe at first. I hope to have a little paper ready to show to Rutherford before I leave, so I'm busy, so busy; but the unbelievable heat here in Manchester doesn't exactly help my diligence. How I look forward to talking to you!

Five days later Bohr had shown his paper to Rutherford and received hearty encouragement to continue. He left Manchester July 24, 1912, his postdoctoral year concluded. His work, however, had only just begun. As he headed home to Denmark, his thoughts turned to a future filled with even greater possibilities than when he had arrived in England the year before.

CHAPTER 2 NOTES

p. 18   "Oh Harald!" Niels Bohr, *Collected Works,* vol. 1, p. 519.

p. 18   "I wonder what he will say . . ." Quoted in Ruth Moore, *Niels Bohr: The Man, His Science, and the World They Changed*, p. 32.

p. 18   "I'm longing to hear . . ." Quoted in Ruth Moore, p. 33.

p. 19   " . . . and then I went on the loveliest . . ." Quoted by Léon Rosenfeld and Erik Rüdinger in "The Decisive Years," in Rozental, ed., p. 44.

p. 19   "The whole thing was very interesting . . ." Niels Bohr, in an interview with T. S. Kuhn, November 7, 1962, Niels Bohr Archives, as quoted by Abraham Pais, *Niels Bohr's Times: In Physics, Philosophy, and Polity*, p. 121.

p. 21   "Among so many scientific bugs . . ." Quoted by Barbara Lovett Cline, *Men Who Made a New Physics*, p. 11.

p. 21   "The crocodile cannot turn . . ." Quoted by Cline, pp. 246–7.

p. 21   "They play games with their symbols . . ." Quoted by Freeman Dyson, *Infinite in All Directions* (New York: Harper and Row, 1988), p. 41.

p. 22   "With one movement from afar . . ." Quoted by Cline, p. 13.

p. 22   "the tremendous . . . argon gas!" Soddy, Frederick, *Atomic Transmutation*, quoted by Rhodes, p. 43.

p. 23   "It was . . . back and hit you." Quoted by Ruth Moore, p. 36.

p. 23   "I know . . . looks like." Quoted by Moore, p. 37.

p. 24   "My name is . . . now recognize." Quoted by Abraham Pais, *Niels Bohr's Times: In Physics, Philosophy, and Polity*, pp. 124–125.

p. 25   "should be glad . . ." Bohr, in an interview with Kuhn et al., November 1, 1962, Niels Bohr Archives, as quoted by Pais, p. 125

p. 25   "Bohr's different! . . ." Quoted by Rosenfeld and Rüdinger in Rozental, ed., p. 46.

p. 25   "on not finding it . . ." C. P. Snow, *The Physicists*, p. 56.

p. 26   "the serious problems of physics . . ." Otto Frisch in *What Little I Remember*, p. 94.

p. 26   "Although there was no doubt . . ." E. N. da C. Andrade, in *The Collected Papers of Rutherford,* vol. 2, p. 299, as quoted by Pais, p. 129.

p. 26–27   " . . . Rutherford is a man you can rely on . . ." Quoted by Rosenfeld and Rüdinger in Rozental, ed., p. 46.

p. 27   "the gloomy cellar . . . at his apparatus." From A. S. Eve, *Rutherford,* p. 239; quoted in Richard Rhodes, *The Making of the Atomic Bomb,* p. 46.

p. 28   "Rutherford . . . thought . . ." Quoted in Rhodes, pp. 68–69 from an oral history interview, American Institute of Physics, New York, p.13.

p. 29   "It seemed to me that . . ." Letter of June 12, Bohr, *Collected Works,* vol. 1, p. 555.

p. 29   "It could be that I've . . ." Bohr, *Collected Works,* vol. 1, p. 559.

p. 30   "I believe I have found . . ." Bohr, *Collected Works,* vol. 1, p. 561.

3

NIELS BOHR, FATHER OF THE ATOM 1913–1924

In the town of Slagelse, about 50 miles southwest of Copenhagen on the island of Sjaelland, Niels Bohr and Margrethe Nørlund were married in a civil ceremony at the town hall on August 1, 1912.

Niels had quietly resigned his membership in the Lutheran Church the previous April. Although he had sought out religion as a child, by the time of their marriage he no longer "was taken" by it, as he put it. "And for me it was exactly the same," Margrethe later explained. "[Interest in religion] disappeared completely," although at the time of their wedding, she was still a member of the Lutheran Church. (Niels's parents were also married in a civil, not a religious, ceremony, and Harald also resigned his membership in the Lutheran Church just before his wedding, a few years later.)

Flags lined the streets of the town on the day of their wedding, and the Nørlunds had made many preparations for a gala celebration, but the ceremony itself was only two minutes long, performed by the chief of police, before members of the immediate family. Harald was the best man. When Niels heard that Margrethe's mother had planned a three-hour wedding dinner, he exclaimed, "How is it really possible to take three hours for a dinner? Can't we take the ferry at 7:00?"

The ferry Niels was so eager to catch would cross the Store Baelt from the island of Sjaelland to the Danish mainland to start the new couple out on the first leg of their honeymoon trip to England, where they stopped off in Cambridge. There Bohr completed his paper on the absorption of alpha particles. When he reached Manchester on August 12, he handed Rutherford his manuscript for publication in *Philosophical Magazine.* In Manchester, Margrethe met Ernest and

Mary Rutherford, who were nearly as enchanted by her as Niels was, delighted that he had found someone so perfectly his match. The four became great friends.

Margrethe and Niels returned to Denmark in September, where they settled into their new life together in Copenhagen. Bohr carried the whirlwind style of his childhood and youth into his adult years. As he had written to Harald from Manchester, "I have so many things that I should like to try . . ." One longtime friend, Jens Rud Nielsen, looking back on that period from 1912 to 1913, recalled, "He would come into the yard, pushing his bicycle, faster than anybody else. He was an incessant worker and seemed always to be in a hurry."

As far as prospects for an academic career were concerned, Bohr clearly realized that, even with a fine record, he had to apply considerable energy if he hoped to have any position at all in his homeland, in a field as esoteric as physics. Denmark, after all, was a small country, and at that time the University of Copenhagen was its only university. Only one professorship in physics existed, and that one wasn't open.

Fortunately, he had thought ahead, knowing he might need several tries to get his foot in the door. Shortly after he received his doctorate, in the summer of 1911, he applied for a docentship (a much lower paid position) in physics. He was turned down. Then, during his stay in England, he heard that his former professor, Christian Christiansen, was resigning, effective August 31, 1912. Bohr applied, even though he knew that the position was much more likely to go to Martin Knudsen, a docent who had far greater seniority. Again Bohr was turned down. Knudsen received the appointment and recommended his own assistant as docent. Now Knudsen's professorial position would not be open for many years—until 1941, as it turned out—and the docentship he vacated was also filled. However, Knudsen offered Bohr a teaching assistantship beginning in the fall of 1912, which Bohr accepted. It wasn't his first, or even second, choice, but it was a start.

Meanwhile, Bohr turned his attention back to the problems that had so intrigued him just before the end of his postdoctoral year: the dilemma of Rutherford's atomic model.

As early as his doctoral dissertation, Bohr had recognized, as he put it, that "One must assume that there are forces in nature of a kind completely different from the usual mechanical sort." By "the usual mechanical sort," he meant forces that classical physics had talked about since the time of Galileo and Newton. In classical terms, energy and matter can be thought of as moving along a ramp, in continuous gradations, with a clear relationship between cause and effect. But very recently, different forces had come to the attention of physicists through the work of two men, Max Planck and Albert Einstein. Their ideas were the stuff of which revolutions are made.

Max Planck didn't seem like the revolutionary type, though. Tall and spare, quiet and dignified, even a bit stuffy and pedantic, he was completely devoted to tradition and authority, both in his physics and in his life. Born in Kiel, Germany in 1858, Planck had specialized in thermodynamics, and he secured an appointment at the University of Berlin in 1889. In general, he appeared to travel pretty much on the beaten path. In fact, when he first started his scientific work in college, one of his professors warned him against pursuing physics. The field, he said, was heading for a dead end. All the great work had already been done, leaving only a few minor details to clear up. But Max Planck was a detail man.

It so happened that one of the "minor details" left to tidy up in the field of thermodynamics was known as the ultraviolet catastrophe, a name dramatic enough to attract anyone's attention. It was the result of problems involved in trying to understand a phenomenon known as black body radiation.

A black body, in physics, is one that absorbs all frequencies of light—something like a piece of coal, only blacker. However, any object with a temperature higher than the temperature of its surroundings loses heat by radiation (the emission of waves or particles through substances). The hotter the object, the more radiation it produces. Logically, since a black body absorbs all frequencies, it should, when heated, radiate all frequencies equally. But that doesn't happen. Instead, black bodies emit larger quantities of some wavelengths than others. When physicists tried to explain these results quantitatively, they came up with an equation that seemed to work well except that it predicted an infinite amount of radiation at the ultraviolet end of the spectrum, which would be impossible, according to the laws of physics. No one could come up with a way to resolve the dilemma in terms of the current physical theory of the 1890s, although several people certainly had tried.

Beginning in 1897, Planck spent the next three years of his life trying to find a solution to this problem. Finally he came up with an idea just before a meeting of the Berlin Physical Society on October 19, 1900. He announced his finding there, and that night colleagues from the meeting rushed to compare the figures to the values found in experiments. They matched—which was very exciting news.

What Planck had proposed was that energy is not infinitely indivisible. Like matter, he said, energy exists not in a continuum, but in discrete, tiny particles or packets. In other words, Planck solved the black body radiation problem by proposing that vibrating particles can radiate only at certain energies—not on a continuum, as classical physics would expect. He said the permitted values could be found by applying a constant, "a universal constant," he said, "which I called h. Since it had the dimension of action (energy $\times$ time), I gave it the name, *elementary quantum of action*." (He used the Latin word *quantum*, an adjective meaning "how much," to describe this new discrete quantity. "Quanta" is the plural form.) In radiation, he maintained, only discrete energies could appear, limited to whole-number multiples of hv; that is, the frequency (v) times Planck's new constant, h. Today this universal constant is known as Planck's constant. Planck calculated it to be a very small number and came amazingly close to the modern calculated value of 6.63 times 10^{-27} erg-seconds. Today Planck's constant is recognized as one of the fundamental constants of the universe.

Because the size of these quanta was in direct proportion to the frequency, radiation at low frequencies requires only small packets or quanta of energy. But, for a frequency twice as high, radiation would require twice the amount of energy.

Following Planck's idea that energy can be emitted only in whole quanta, it becomes very easy for a body to radiate at low frequencies—not that much energy has to be pulled together to make up a quantum of energy. But at high frequencies, pulling a quantum's-worth of energy together is not so easy. The quantum-energy requirements to radiate at the high-frequency end of the spectrum are so great that it's very unlikely to happen. So black bodies do not radiate all frequencies equally, and that's the key to the "ultraviolet catastrophe."

Conservative Max Planck had no heart for pushing this idea to its natural, radical conclusions, but another, younger physicist made use of Planck's theory in a paper, published in 1905, that would win him the Nobel Prize in physics in 1921. That young physicist was Albert Einstein.

If Max Planck was noted for his middle-class conservatism, Albert Einstein was the epitome of the complete rebel, a loner who preferred to work alone, a wanderer in the highest realms of thought, uncomfortable with and often disdainful of the everyday preoccupations of the average man or woman. As he once explained, he sought out science to get away from the "I" and the "we." He preferred to think instead about the "it." Supremely confident of his own genius, like his greatest predecessor, Isaac Newton, Einstein walked his own way, to his own challenges, and worried little about what he called "the chains of the merely personal . . . dominated by wishes, hopes, and primitive feelings."

By 1905, at the age of 26, he had produced no fewer than five papers, all of which were published that year by the German *Yearbook of Physics.* In the first he examined a topic that would ultimately win him the Nobel Prize. In the second, he introduced what would later be called the special theory of relativity, one of his two most famous contributions to physics. In the special theory of relativity, Einstein established his famous formula $E=mc^2$, where E is energy, m is mass and c represents the speed of light, which is always constant. This formula showed the interrelationship of mass and energy, and by using it, Einstein could show that the laws of physics are the same in all reference frames that are moving at constant velocities relative to one another. (He later succeeded in applying his theory of relativity to the more general case of accelerated systems, establishing a new theory of gravitation of which Newton's classic theory was a special case. The general theory of relativity, published in 1915, held enormous implications for understanding how the universe works.) In a third paper published in 1905, he examined an aspect of statistical mechanics and Brownian motion, the apparently erratic movement of pollen in fluids. Another two explored further aspects of special theory of relativity and Brownian motion.

But his Nobel-prize-winning work dealt with a mysterious phenomenon known as the photoelectric effect, that is, the fact that when light falls on certain metals, electrons are emitted. (It's the principle that makes solar arrays on satellites and spacecraft work.) But, strangely, there is no connection between the brightness of the light and its ability to knock electrons free of the metal. Instead, it is the *color* of the light that matters—its frequency. No one had been able to explain this mystifying fact. Classical physics—that is, explanations based on what Bohr would call "forces in nature of the usual mechanical sort"—could offer no explanation.

That's where Einstein stepped in, making use of Planck's quantum theory, which had been gathering dust for a couple of years without too much attention. Einstein pointed out that a particular wavelength of light is made up of quanta of fixed energy content, according to quantum theory. When a quantum of energy bombards an atom of a metal, the atom releases an electron of fixed energy content and no other. A brighter light would contain more quanta, still always of fixed energy content, causing the emission of more electrons, also still all of the same energy content. When the light's wavelength is shorter (and the frequency higher), more energy is contained in the quanta and the electrons released are more energetic. Very long wavelengths (of lower frequency) would be made up of quanta having much smaller energy content, in some cases too small to cause any electrons to be released. And this threshold would vary depending on the metal.

Once again Planck's theory had succeeded in explaining a physical phenomenon where classical physics could not. It was the first major step in establishing what would become known as quantum mechanics, the recognition of the discrete and discontinuous nature of all matter, especially noticeable on the scale of the very small.

Bohr's great insight was to see how the theory of the quantum could be used to explain how things work within the structure of atoms and their particles—specifically to overcome the problem of mechanical instability in Rutherford's model. As he left Manchester in July, he had already arrived at the kernel of his thinking, namely that since classical mechanics predicted instability where there clearly was none, he would call on the quantum approach to describe what actually occurred. Planck had first introduced quantum principles to explain what could not be explained classically in thermodynamics.

Einstein had done the same for light. Now Bohr planned to test out quantum principles as a way of explaining what went on within the atom itself.

Once back in Denmark, Bohr worked on these ideas all during the fall and early winter months of 1912. First, he assumed that electrons could not orbit an atom's nucleus willy-nilly in just any orbit. He proposed that "stationary states," as he called them, must exist in the atom—specific orbits that electrons could occupy without spiraling inward and crashing into the nucleus. When he figured the numbers for his model, he found that they coincided with several experimental results. When scientists see this sort of coincidence—as Max Planck's colleagues also had—they recognize that it can be an indication that they are heading down the right path. Of course, it also can be sheer coincidence. What bothered Bohr about his model was that it was arbitrary. It explained some chemical phenomena, but it seemed no more real that Thomson's plum pudding model.

Then Bohr happened to come across a series of papers by J. W. Nicholson, a professor of mathematics at King's College, London. Bohr had met Nicholson and didn't think much of his abilities, but what startled him was that Nicholson had proposed an explanation for an unusual spectrum in the corona of the Sun by suggesting a model of the atom that used quantum principles. Nicholson had made mistakes, but Bohr began to worry about competition. He had hold of such an exciting idea, and he hoped he would have time to work it out in detail and publish before someone else got there. So Nicholson's paper threw more fuel on an already intensely burning flame. More important, however, Nicholson's introduction of spectra into the picture gave Bohr an idea.

He had not even considered the idea that the spectra produced by the elements might be related to his problem. In the last interview he gave, he said, "The spectra was a very difficult problem. . . . One thought that this is marvelous, but it is not possible to make progress there. Just as if you have the wing of a butterfly, then certainly it is very regular with the colors and so on, but nobody thought that one could get the basis of biology from the coloring of the wing of a butterfly." And so, who would think that the coloring of spectra would have anything to do with the basis of physics?

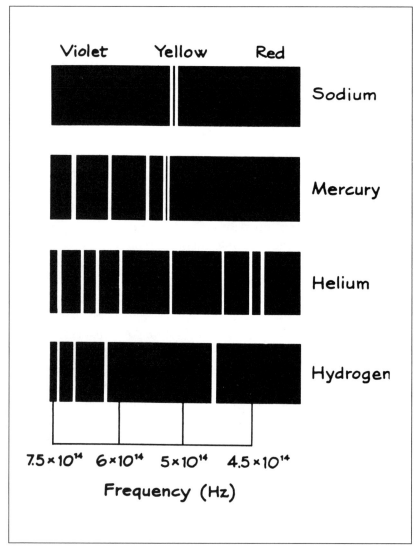

Figure 4. Emission spectra for sodium, mercury, helium, and hydrogen. Each element displays a unique discrete spectrum, almost like a finger-print, a fact that puzzled scientists for years until Bohr came up with his model of the atom. Bohr suggested that electrons release energy only while making jumps from one orbit to another (explaining the discrete spectral lines emitted by the elements) and that the electrons move only in certain allowed orbits at specific distances from the nucleus of each element. (From Jacqueline D. Spears and Dean Zollman, *The Fascination of Physics.* Benjamin/Cummings, 1985.)

In the 19th century physicists had discovered that each element produces a characteristic spectrum of light when heated. Sodium, for instance, emits light only at particular wavelengths—yellow, in this case. Potassium emits a violet light. And so on. In terms of Planck's theorem, that meant that the atoms of each element produce light quanta only of a particular energy. Physicists had observed that certain spectra were associated with the atoms of certain elements, but they had never been able to explain why.

Some had found some fascinating mathematical regularities, however. In 1885, Johann Balmer, a mathematical physicist from Switzerland, had found a formula for calculating the wavelengths of the spectral lines of hydrogen. Five years later Johannes Rydberg, a Swedish spectroscopist (a specialist in producing and studying spectral lines), came up with a general formula that could be used to calculate several different line spectra. But these mathematical harmonies served only to deepen the mystery; they offered no explanation.

With his model of the atom, Bohr set out to explain why. He realized that the radius of an electron's orbit was determined by Planck's constant. That meant that the amount of energy also was fixed. An electron in a permissible orbit did not emit any radiation; normally it occupied a stable, basic orbit that Bohr called a ground state.

But if energy was added to the atom, if it was heated, for example, the electron would respond by jumping from one orbit to another—and when it did that, it changed energy states. In response to heat, it would jump to a higher orbit, one farther from the nucleus. If the heat was turned higher, the electron would keep jumping to higher and higher orbits. If it was cooled off, the electron would jump inward to a lower orbit, closer to the nucleus. Moving to a larger orbit, it would absorb energy; moving to a smaller one, it would emit energy. When energy no longer was added, the electron would return to its ground state.

Bohr did the calculations for hydrogen's single electron. He worked out the energies involved for jumping from one orbit that was permissible to another. He calculated the light wavelengths that would be produced if the energy was converted to light. It worked! His calculations matched the spectrum of hydrogen, which always before had been a mystery.

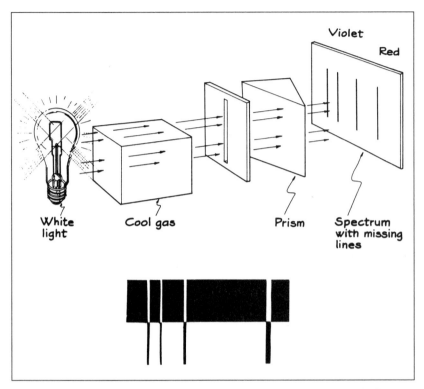

Figure 5. In the top illustration, light emitted by an incandescent bulb produces an absorption spectrum when it passes through a sample of cool gas before striking a screen. The gas has absorbed energy from the white light of the bulb, as indicated by the dark lines on the screen. From the pattern of dark lines, it is possible to tell which frequencies the gas has absorbed. Because every element has a unique spectrum, the gas can be identified from the pattern. When an element is heated, it gives off (emits) energy that can be displayed in the same way, producing a similar emission spectrum—a negative, or inversion, of the absorption spectrum. The bottom illustration shows the emission and absorption spectra of hydrogen. (From Jacqueline D. Spears and Dean Zollman, *The Fascination of Physics.* Benjamin/Cummings, 1985.)

This is what he came up with: Each electron emits a photon (a quantum of light) of characteristic energy, or frequency. The jumps—and the photon energies—are limited by Planck's constant. If the lower-energy state (when the atom is cool and the electron is close to the nucleus) is called W^1, and the higher-energy state is called W^2, you can subtract W^2 from W^1 and the result is hv (Planck's constant, h, times the frequency, v). Bohr saw this and, using the formula $W^2 - W^1 = hv$, he was able to come up with Balmer's series. He also found that with a slightly more complex formula, using nonarbitrary numbers, representing the mass of the electron, its charge, and Planck's constant, he could come up with Rydberg's experimental values!

As always, Bohr had great difficulty bringing his paper to conclusion. It was getting too long for publication in any journal. Finally he decided to divide it up into three parts. (For this reason, it usually is referred to as the trilogy.) After Bohr had finally finished Part I, "On the Constitution of Atoms and Molecules," he anxiously sent it off to Rutherford on March 6, 1913, and held his breath. In the meantime, he continued to write and rewrite the other two parts.

When Rutherford's letter, dated March 20, finally arrived, Bohr tore it open and scanned it quickly. With the first sentence, he breathed a great sigh of relief.

"I have received your paper safely and read it with great interest," wrote Rutherford, "but I want to look over it again carefully when I have more leisure. Your ideas as to the mode of origin of the spectrum of hydrogen are very ingenious and seem to work out well." But Rutherford's role was to critique and so he did: ". . . but the mixture of Planck's ideas with the old mechanics make it very difficult to form a physical idea of what is the basis of it." And a bit later, he raised a thorny question in a characteristically direct way:

> There appears to me one grave difficulty in your hypothesis, which I have no doubt you fully realize, namely, how does an electron decide what frequency it is going to vibrate at when it passes from one stationary state to the other? It seems to me that you would have to assume that the electron knows beforehand where it is going to stop.

Overall, though, Rutherford thought the paper should go off to the *Philosophical Magazine.* He offered to clean up the English for Bohr, and, since he thought it was much too long and wordy, he concluded somewhat casually with the comment, "I suppose you have no objection to my using my judgment to cut out any matter I may consider unnecessary in your paper? Please reply."

Bohr didn't know what to say in reply. He had labored over every sentence, every word of the paper, and was certain that nothing could be cut out without damaging the meaning drastically. But how could he convey a "hands-off" message to Rutherford tactfully? Finally he wrote back that he would be "glad for any alterations you consider suitable," apologizing for any trouble he might have caused, and then concluded offhandedly, "I now have some holidays and I have decided to come over to Manchester." It was a totally unplanned trip, in fact, but it was the only way Bohr could see out of his dilemma.

When Bohr reached Manchester, he was greeted warmly and the two sat right down to work, Rutherford with the idea that at least one-third of the manuscript could be jettisoned. Bohr tenaciously defended the existence of every sentence. Section by section they went through the paper. Section by section Rutherford was won over. In the end, he made only a few corrections to Bohr's English and sent the paper off intact. Later, he loved to tell the story of Bohr's bulldog defense of his words and how he had finally had to give in to the gentle if tenacious Dane.

The second part in Bohr's 1913 trilogy bore the title "Systems Containing Only a Single Nucleus," and the third, "Systems Containing Several Nuclei." All three parts were published by the end of 1913, and Bohr's trilogy became seminally important to physics. It not only offered a highly useful model of the atom, but it showed that quantum mechanics was a fundamental part of how nature worked. While the mechanistic physics of Newton had worked well on larger scales, it could not account for subtleties on the atomic scale. For this work, Bohr would win the 1922 Nobel Prize in physics.

Bohr knew that his model was only a very sketchy approximation of reality, however. Otto Frisch would later recall, "Bohr himself was very much aware of the crudeness of that model; it resembled the atom no more than a quick pencil sketch resembles a living human face. But he also knew how profoundly difficult it would be to get a better picture."

Late in his life Bohr would reflect, "It was clear, and that was *the* point about the Rutherford atom, that we had something from which we could not proceed at all in any other way than by radical change. And that was the reason then that [I] took it up so seriously."

It was radical. It is the threshold of what became known as the heroic age of quantum physics. And the reaction was enormous. Frisch would later recall, "That picture was so astonishing and

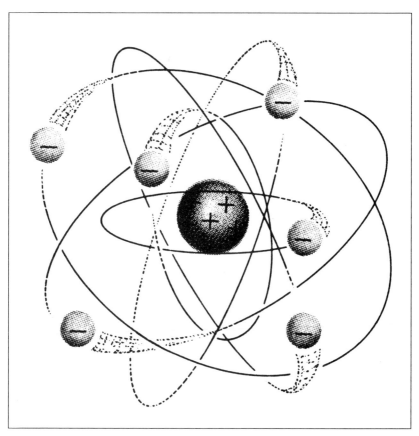

Figure 6. In 1913 Niels Bohr proposed a model of the atom with electrons confined to specific circular orbits around the nucleus. (From G. Tyler Miller Jr., *Chemistry: A Basic Introduction,* 2nd ed. Belmont, Calif.: Wadsworth Publishing Company, 1981.)

unorthodox at the time that a number of physicists . . . had sworn to give up physics if that nonsense was true (none of them did)."

Rutherford himself continued to be cautious. In March 1914 he said "While it is too early to say whether the theories of Bohr are valid, his contributions . . . are of great importance and interest." And in August, "N. Bohr has faced the difficulties by bringing in the idea of the quantum. At all events there is something going on which is inexplicable by the older mechanics."

The old guard was unlikely to be ecstatic. Lord Rayleigh, past 70 by that time, was uninterested. J. J. Thomson did not even comment on Bohr's 1913 papers until 1936.

Even Einstein, a young radical himself, was lukewarm at first. Hevesy told him the news in September 1913 in Vienna, to which Einstein replied that Bohr's work seemed very interesting and important if right—a polite way of expressing skepticism. Then Hevesy told him about the spectral lines, and Einstein brightened. "This is an enormous achievement," he said. "The theory of Bohr must then be right."

Many years later, when Einstein was almost 70, he wrote:

> That this insecure and contradictory foundation [of physics in the years from 1910 to 1920] was sufficient to enable a man of Bohr's unique instinct and tact to discover the major laws of the spectral lines and of the electron shells of the atoms together with their significance for chemistry appeared to me like a miracle—and appears to me as a miracle even today. This is the highest form of musicality in the sphere of thought.

Einstein and Bohr would not meet for another seven years, but during the years that followed publication of Bohr's 1913 trilogy, Einstein kept track of Bohr's work through his writings. He also gathered news from a young physicist named Paul Ehrenfest, of Amsterdam, who struck up a friendship with Bohr by letter in 1918. In a postcard to Max Planck in 1919, Einstein wrote, "Ehrenfest tells me many details from Niels Bohr's *Gedankenküche* [thought kitchen]; his must be a first-rate mind, extremely critical and far-seeing, which never loses track of the grand design."

Harald wrote from Göttingen that people were reading Niels's work with interest, but many seemed to think his premises were too fantastic and too bold. Bohr spoke at a meeting in Birmingham, England on September 12, 1913, the first time he presented his work at an international meeting. The write-up in the journal *Nature* said, "Dr. Bohr . . . arrived at a convincing and brilliant explanation of the laws of spectral series."

By the winter of 1914, the University of Copenhagen, impressed by now, invited him to submit his application for appointment as professor of theoretical physics. It would be the first time that theoretical physics was taught at Copenhagen as a separate science, and it was exactly what he wanted. Bohr submitted his application March 4 and wrote to Rutherford, asking for a recommendation, which Rutherford sent back, filled with praise, by return mail. Approval by the faculty was almost unanimous.

But, before the appointment was actually in place, Bohr received another letter from Rutherford that changed everything. An appointment of a two-year readership was available in Manchester, with a stipend of 200 pounds. Rutherford wrote, "I should like to get a young fellow with some originality in him." Namely, Bohr.

Premonitions of war were brewing in Germany at this time, but for the most part, people thought war was unlikely, and Bohr couldn't pass up an opportunity to work directly with Rutherford. He quickly arranged with the University of Copenhagen to wait two years for him and made preparations to head back to Manchester.

On June 28, 1914 the Archduke Francis Ferdinand of Austria-Hungary was assassinated by a Serb extremist. Austria-Hungary issued an ultimatum to the Serbian government, and Germany promised support. Tensions that had been seething for years suddenly sent the world to the brink of war. Russia mobilized against Austria, and on August 1, Germany declared war on Russia. Denmark, as always, was vulnerable. German and British fleets prowled the waters, and one sea battle took place on August 28. Travel to Manchester now looked risky.

But by September, although war in Europe continued, travel between Denmark and England by sea again seemed safe, and the Bohrs set out for Manchester. There Bohr produced another paper for the *Philosophical Magazine* in 1915, "On the Quantum Theory of Radiation and the Structure of the Atom," strengthening the points of his previous work. As before, the work in Manchester was exhilarating, intense, and always productive. To Bohr, it was an object lesson in how he might run his own department of theoretical physics in Copenhagen, when he returned.

By 1916, as Bohr was preparing his notes and papers to return to Denmark, the war had grown grimmer. Rutherford sent a letter with Bohr, to protect his notes from confiscation, certifying that they were intended for English scientific publication. Although the guns in Europe roared on, the Bohrs returned safely home.

Papers by other scientists were beginning to come out that were offshoots of Bohr's theory, and he was having to work hard to keep ahead. Three years had passed since his famous trilogy had been published, and confirming experimental results were beginning to pile up. So were the offers. Before the year was out, Bohr had received an invitation from the University of California to spend a year there. Soon after, the University of Manchester invited him to join its faculty. But the war made California an unlikely destination, and Bohr asked Manchester to wait a few months for his decision. Meanwhile, in November 1916, the Bohrs had their first son, Christian, an event they both rejoiced in greatly. Five other children, all boys, would follow in the years to come: Hans Henrik (born in 1918), Erik (1920), Aage Niels (1922), Ernest David (1924, named after both their good friend Rutherford and Niels's maternal grandfather), and Harald (1928). Bohr loved his children and always managed to spend time with them, despite his fast-paced schedule.

Throughout 1916, Bohr continued to work on publications, finding them difficult to complete, as always, and he continued an active correspondence with Rutherford, the news of whose research he followed with great interest. But now his attention was also greatly drawn to his teaching. At first his classes in theoretical physics drew only a handful of advanced students, primarily those interested in quantum theory. (Knudsen, the only other physics professor at the university, had no use for quantum theory. One student who took classes from both of them recalled Knudsen's sharp response to a question involving it: "If we have to use quantum theory to explain this, we may as well not explain it.") Gradually, as Bohr delved more and more into atomic theory in his classes, faculty members began turning up too. The discussions went deep and Bohr pushed his students' thinking, chalk in hand, furrowed brow, expectant smile, waiting for the next thought. His was a kindly, responsive face. Abraham Pais, a physicist from Holland who later became his biographer, once wrote of his first meeting with Bohr, "My first thought was, what a gloomy face." But this first impression melted immediately. Those who knew Bohr were captured by the sunny, warm smile and the animation that came to his face when he talked.

By 1917 Bohr had approached the university about an idea he had—the establishment of a small institute for theoretical physics as part of the university. Soon after the war's end, the plan was approved. Bohr was able to raise about $20,000 toward the construction of a building, and Copenhagen made a site available adjacent to a park near the center of the city. In his enthusiasm, Bohr immediately invited the Rutherfords to the inauguration, which was still several years off.

All these plans nearly toppled, however, when a letter arrived from Rutherford. It had been delayed by the disruption of the mails caused by the confusion as the war ended. It was marked "private and confidential." Rutherford was offering Bohr a permanent position as professor of mathematical physics at a new center for modern physics research at Manchester. He wrote:

> You know how delighted we would be to see you working with us again. I think the two of us could try and make physics boom. Well think it over and let me know your mind as soon as you can. Possibly you might think of visiting us as soon as the seas are clear.

Rutherford concluded his letter, "I wish I had you here to discuss the meaning of some of my results on collisions of nuclei. I think I have got some rather startling results."

This last was Rutherford's strongest card. What a team they might make! It was an opportunity without parallel—wonderful and impossible at the same time. Niels and Margrethe hurried to discuss the new turn of events with Harald. They talked all day and into the night. In the end, though, there was no question. Bohr was a Dane, and he could not turn his back on Denmark, when the country had offered him everything he had asked for. He had to stay and build his own center for modern physics research, in Copenhagen. Bohr felt, at that moment, that he might not be making the best decision either for his scientific work or his family's financial welfare. Working with Rutherford, he might accomplish insights he might never have on his own, and Denmark could not offer him the salary that the English university could. But he made the only ethical decision he could make. Rutherford did not give up immediately, but in the end, Bohr stayed to build an institute for Denmark.

And build he did, from the ground up. Every detail of the architect's drawings had to meet with his approval. He went to the site every day to oversee the progress during construction, and he watch-dogged at

Niels Bohr in the 1920s.   (Niels Bohr Archive, Copenhagen)

every turn. He may have driven the architect and construction crews crazy, and he certainly slowed the progress, but in the end the institute was built exactly to his specifications.

Rutherford, meanwhile, was offered the directorship of the Cavendish Laboratory at Cambridge, which he accepted. Bohr took time off in July 1919 to visit him and consult with him about his work. He also lectured in Holland, where he and other physicists talked about atomic problems nonstop. Arnold Sommerfeld, the first of the German scientists to venture into the previously neutral countries, went to Copenhagen, where he and Bohr talked about quantum theory and atomic structure.

For a time the institute ran out of construction funds, as costs raced sky-high in the postwar economy. Always resourceful, Bohr obtained a grant from the Carlsberg Beer Company in Copenhagen, and construction resumed.

But, despite his busy schedule, when Max Planck invited him to go to Berlin to lecture on spectral theory at the *Physikalische Gesellschaft* (Physics Society), Bohr quickly agreed, and it was a decision he never regretted. There, as he arrived at the Berlin physics building, Planck and Albert Einstein came out to greet him. Planck looked formal and proper, as he smiled warmly behind wire-rimmed glasses. Einstein, with his wild halo of hair, never looked formal, on that day or any other day. The three men began discussing atomic physics immediately and continued, morning to night, for the rest of the conference, whenever they weren't in meetings.

Einstein had formulated the general statistical rules for the electron jumps from one of Bohr's stationary states to another. These jumps could occur, he said, not just in response to radiation or collision. They could occur spontaneously.

Bohr, meanwhile, suggested in his lecture on spectral theory that an exact determination of where and when electrons jump could not be made. This Einstein did not like. No theory should leave to chance the time and determination of fundamental processes, he maintained. Here, already, the two great minds began to take divergent paths. Their conversations revolved around these themes and left a deep impression on Bohr, as he continued to mull them over in his thoughts.

"With his mastery for coordinating apparently contrasting experience," Bohr later wrote, "without abandoning continuity and causality, Einstein was perhaps more reluctant to renounce such ideals than someone for whom renunciation . . . appeared to be the only way to proceed with the immediate task of coordinating the multifarious evidence regarding atomic phenomena."

Two days after the end of the conference, Einstein wrote to Ehrenfest, "Bohr was here, and I am as much in love with him as you are. He is like an extremely sensitive child who moves around in this world in a sort of trance."

And to Einstein, Bohr wrote, "To meet you and talk with you was one of the greatest experiences I ever had."

In September 1921 the inauguration of the Institute for Theoretical Physics in Copenhagen took place, and Bohr became its head at the age of 36. From the time the institute was built, the Bohrs lived in a flat on the upper level of the research center's only building. In 1924 Bohr bought a summer home as a retreat in Tisvilde on the island of Sjaelland (where Copenhagen is located), 40 miles northwest of the Danish capital, near the shore of the Kattegat strait, which lies between Denmark and Sweden. An old gamekeeper's home, the house was a one-story cottage with a thatched roof. Its name was *Lynghuset,* meaning "heather house," and it was nestled among rolling hills dotted with heather shrubs and high pine forests. A small one-room cabin, which they called *Pavillonen* (the pavilion), stood nearby the house—the perfect place for Bohr and his colleagues to work undisturbed. The peaceful surroundings of Tisvilde provided the Bohrs with the ideal summer retreat, a much-needed place to get away from the pressures of their intense daily life.

But from the start Bohr's life revolved around the new institute. There, like a magnet, Bohr drew the best young minds from all over the world. Georg von Hevesy, from his days at Manchester, was one of the first of many foreigners Bohr invited to come to Denmark to work in the new institute. Fueled with the exciting discoveries of radioactivity, quantum theory, and relativity, in its first 25 or 30 years the 20th century witnessed an enormous outpouring of ideas and discoveries unparalleled in the history of physics. A dynamic cluster of men and women—ambitious, brilliant, keenly prepared, and talented—gathered in the universities of Europe, Britain, and, to a lesser degree, Canada and the United States to ride the crest of a great wave of exploration into the inner regions of the atom. Niels Bohr's institute

The Niels Bohr Institute at the time of its inauguration in 1921. (Niels Bohr Archive, Copenhagen)

rapidly became a point of reference for all these young physicists, led by the man who came to be known as the "Gentle Dane."

Niels's easygoing sense of humor and charismatic way of drawing ideas out of people stimulated endless discussion. As Otto Frisch, who was a young student at the institute, described him.

> He had a soft voice with a Danish accent, and we were not always sure whether he was speaking English or German; he spoke both with equal ease and kept switching. Here, I felt, was Socrates come to life, tossing us challenges in his gentle way, lifting each argument to a higher plane, drawing wisdom out of us which we didn't know we had, and which of course we hadn't.

As a teacher and mentor, he was unrivaled.

CHAPTER 3 NOTES

p. 33   "was taken . . . completely." From an interview with Margrethe Bohr, her son Aage, and Léon Rosenfeld by Thomas S. Kuhn, January 30, 1963, Niels Bohr Archives, quoted by

Abraham Pais, *Niels Bohr's Times: In Physics, Philosophy, and Polity,* p. 134.

p. 33　"How is it really . . ." Margrethe Bohr, from interview by Kuhn, quoted by Pais, p. 134.

p. 34　"I have so many . . ." Quoted by Ruth Moore, *Niels Bohr: The Man, His Science, and the World They Changed,* p. 42.

p. 34　"He would come . . ." J. Rud Nielson, *Physics Today* (October 1963), p. 22.

p. 34　"One must assume . . ." From J. L. Heilbron and Thomas S. Kuhn, "The Genesis of the Bohr Atom," *Historical Studies in the Physical Sciences,* vol. 1 (1969), p. 214, quoted by Richard Rhodes, *The Making of the Atomic Bomb,* p. 70.

p. 36　"a universal constant . . ." From Heilbron and Kuhn, p. 256, quoted by Rhodes, p. 70.

p. 37　"The chains . . . feelings." Quoted by Barbara Lovett Cline, *Men Who Made a New Physics,* p. 73.

p. 39　"The spectra . . ." From Heilbron and Kuhn, p. 257, quoted by Rhodes, p. 72.

p. 43　"I have received your paper . . . please reply." Quoted by Ruth Moore, p. 59–60.

p. 44　"I now have . . . Manchester." Quoted by Moore, p. 61.

p. 44　"Bohr himself . . ." Otto Frisch, *What Little I Remember,* p. 93.

p. 44　"It was clear . . ." Niels Bohr, oral history interview, American Institute of Physics, p. 13.

p. 44–45　"That picture . . ." Frisch, p. 93.

p. 45　"While it is too . . . mechanics." Rutherford, quoted by Pais, p. 153.

p. 46　"This is an enormous . . ." G. von Hevesy, letter to Niels Bohr, September 23, 1913, in Niels Bohr, *Collected Works,* vol. 2, p. 532.

p. 46　"That this insecure . . ." Quoted in Pais, *Subtle Is the Lord: The Science and Life of Albert Einstein,* p. 416.

p. 46　"Ehrenfest tells . . ." Quoted in Pais, *Subtle Is the Lord,* p. 416.

p. 46   "Dr. Bohr . . ." *Nature,* 92 (1913), p. 304.
p. 47   "I should . . . originality in him." Quoted by Ruth Moore, p. 73.
p. 48   "If we have to use . . ." Quoted in Ruth Moore, p. 95.
p. 48   "My first thought . . ." Pais, p. 5.
p. 49   "You know how delighted . . ." Quoted in Ruth Moore, p. 97.
p. 51–52   "With his mastery . . ." Quoted in Ruth Moore, p. 105.
p. 52   "Bohr was here . . ." Pais, pp. 416–17.
p. 52   "To meet you . . ." Pais, p. 417.
p. 53   "He had a soft voice . . ." Frisch, p. 92.

4

THE GENTLE GIANTS, BOHR AND EINSTEIN: AN INTELLECTUAL TUG OF WAR 1925-1929

During the 1920s and 1930s, the Institute for Theoretical Physics in Copenhagen, headed by Bohr, commanded an influence over the world of scientific thought equaled only by Aristotle's Lyceum in Athens. Theoretical physicists went there from all over the world, during a time often called the heroic age of atomic physics. These were the same years—especially 1925 to 1927—that in Göttingen are known as the years of *Knabenphysik*, "boyhood physics." Breakthrough after breakthrough was made by very young scientists: Werner Heisenberg, at 23; Wolfgang Pauli, at 25; Paul Dirac, at 22. At 37, Erwin Schrödinger, by contrast, seemed out of place. All of these found their way to Niels Bohr's institute and felt the influence of Bohr the mentor, himself only in his early 40s. One by one, each carved out a place for himself in the quantum physics hall of fame. The year 1925, in particular, was a great year in physics, especially quantum physics.

Wolfgang Pauli was clumsy in the lab and faltering in front of a lecture hall full of listeners. Yet his mind could pierce to the heart of a problem seemingly without effort. He studied under Arnold Sommerfeld at the University of Munich, where he did his doctoral work, and then pursued postgraduate studies both in Copenhagen with Bohr and at Göttingen. (Later he moved to the United States, joining the Institute for Advanced Study in Princeton, and becoming a citizen in 1946.)

His great insight, known as Pauli's exclusion principle, came to him as he was battling a problem known as the anomalous Zeeman effect. Moody and dejected while visiting Bohr, he replied snappishly to Margrethe Bohr's solicitous inquiry, "Of course I am unhappy! I cannot understand the anomalous Zeeman effect."

Pauli based his work on an enormous pile of data, in which he discerned a simple sorting-out principle that held true in all cases: In any system of elementary particles—for example, the collection of electrons within the atom—no two particles may move in the same way, that is, occupy the same energy state. He announced this "Exclusion Principle" in 1925.

The Exclusion Principle explained why electrons in an atom don't all drop down to the orbit nearest the nucleus—which might normally be expected, since an electron traveling along the smallest, closest ring requires the least amount of energy to complete an orbit. According to Pauli's principle, once an electron is in an orbit, it excludes any other electron from occupying the same orbit. Over the years Pauli's Exclusion Principle has proved to hold true for nuclear particles that no one had even dreamed of, and it has been a key concept in the development of quantum mechanics. Pauli received the Nobel Prize in physics (somewhat belatedly) for this work in 1945.

In Paris, meanwhile, Louis de Broglie proposed an idea, known as the wave-particle duality, that ended up bothering Albert Einstein a great deal. De Broglie maintained that if atomic particles were also thought of as waves, not just particles, the consequences theoretically were very nice, even though the idea is mind-boggling.

According to Planck and Einstein, light, which had most recently been regarded as a wave, should be regarded as a particle. Now de Broglie was saying that particles—photons, electrons, even atoms— sometimes behave like waves. Outlandish as this sounds, experimental tests bear out de Broglie's theory.

The idea immediately caught on among many physicists. In 1926, Erwin Schrödinger came up with an equation that accurately described the way an electron behaves. Schrödinger interpreted his equation, known as "wave mechanics," to represent that electrons *are* de Broglie's waves. Swept up in the wave concept, Schrödinger wanted to abandon completely the idea that electrons are particles.

Erwin Schrödinger. (Photo by Francis Simon, courtesy AIP Emilio Segrè Visual Archives)

But on this point Schrödinger was wrong. Schrödinger's theory of wave mechanics worked beautifully, and physicists loved it for its perfect accuracy, but in June of the same year, Max Born published a paper that instead gave electron waves a probabilistic interpretation.

Born said the rise and fall of waves could be taken to indicate the rise and fall in *probability* that the electron behaved as a particle.

In this brief, basic paper, Born went straight to the crux of the problem of cause and effect. In the macroworld of classical mechanics, cause and effect might work as a principle, but not in the subtle microworld where quantum mechanics ruled. Regarding atomic collisions, Born wrote:

> One does not get an answer to the question, What is the state after collision? but only to the question, How probable is a given effect of the collision? . . . From the standpoint of our quantum mechanics, there is no quantity which causally fixes the effect of a collision in an individual event. Should we hope to discover such properties later . . . and determine [them] in individual events? . . . I myself am inclined to renounce determinism in the atomic world, but that is a philosophical question for which physical arguments alone do not set standards.

Basically, Born had abandoned causality in the classical sense. Leon Lederman, one of the world's top physicists in the 1990s, calls Max Born's interpretation "the single most dramatic and major change in our world view since Newton." Yet Schrödinger wasn't happy with Born's idea at all, and neither were many other classical physicists of his day. Born's probability meant that the determinism promised by Newton's laws of physics could no longer be relied upon. Coupled with quantum theory, it meant that only probabilities could be known—about anything you wanted to measure.

By 1926–27 Albert Einstein, who already had a towering reputation in physics, began to reject many of the bizarre implications of quantum mechanics. Replying to one of Born's letters, Einstein wrote: "Quantum mechanics is very impressive. But an inner voice tells me that it is not yet the real thing. The theory produces a good deal but hardly brings us closer to the secret of the Old One. I am at all events convinced that *He* does not play dice."

In February 1927, Einstein gave a lecture in Berlin, in which he reportedly said, "What nature demands from us is not a quantum theory or a wave theory; rather, nature demands from us a synthesis of these two views which thus far has exceeded the mental powers of physicists."

But Bohr, Sommerfeld, Heisenberg, and others took Born's ideas in stride—the concepts seemed logical—and they and their colleagues continued the exciting work of trying to get all the pieces to fit. It was great teamwork. It was also the beginning of a friendly but impassioned lifelong debate between Bohr and Einstein, which would continue for more than a quarter of a century until Einstein's death in 1955.

The issues heated up in March 1927, when Werner Heisenberg produced another amazing physical theory: the Uncertainty Principle. According to that principle, the exact position and precise velocity of an electron could not be determined at the same time. In other words, no one can tell for certain where an electron will go when it is hit—all that can be said is where it probably will go. Only statistical predictions can be made.

This idea capped off the great scientific revolution we call quantum theory. (Although much remained to be wrapped up and quantum field theory is still evolving today. Some scientists contend that the theory will not be complete until it is fully combined with gravitation, and many have attempted to formulate a unified field theory that would accomplish this goal.) The work of Born, Schrödinger and Heisenberg formed the basis for quantum mechanics, which has been used to interpret chemistry and subatomic physics with great success. Yet Einstein never accepted the Uncertainty Principle, and he debated long and ardently with Bohr about it.

In September 1927 Bohr attended a meeting in Colma, Italy. (Einstein had been invited, but he did not attend.) There, on September 16, Bohr outlined for the first time the principle he called "complementarity":

> The very nature of the quantum theory . . . forces us to regard the space-time coordination and the claim of causality, the union of which characterizes the classical theories, as complementary but exclusive features of the description, symbolizing the idealization of observation and definition, respectively.

Bohr maintained that a phenomenon can be looked upon in two mutually exclusive ways, and yet both outlooks can remain valid in their own terms. For example, you may know the momentum *or* the position of an electron, but not both. As his student Frisch put it, "It is a bit as if reality was painted on both sides of a canvas so that you could only see one aspect of it clearly at any time." Complementarity

Werner Heisenberg. (Photo by Friedrich Hund, courtesy AIP Emilio Segrè Visual Archives)

would become the cornerstone of Bohr's thinking about quantum and classical physics, his way of reconciling two equally plausible but mutually exclusive ideas.

Brilliant, vital, and exciting ideas, the emerging concepts of quantum physics were also unsettling. Even Bohr, having put quantum to its first use in the physics of matter, recognized the theory's enormous mysteries.

"It makes me quite giddy to think about these problems," a visitor once complained to Bohr. "But, but, but . . ." Bohr stammered ingenuously, "if anybody says he can think about quantum theory without getting giddy it merely shows that he hasn't understood the first thing about it!"

As the brilliant American theoretical physicist Richard Feynman used to say to his students, "I think it is safe to say that no one understands quantum mechanics. Do not keep saying to yourself, if you possibly can avoid it, 'But how can it be like that?' because you will go 'down the drain,' into a blind alley from which nobody has yet escaped. Nobody knows how it can be like that."

During the years from 1925 to 1929 Bohr gave many lectures on this complex topic, and his listeners valiantly strove to follow along. But not everyone was successful at keeping up. As Robert Courant recounted, at one point during the "grand period of the evolving quantum theory" Bohr was laid up from a skiing accident, with only Harald and Courant to listen to him. Bohr began trying to explain to them the "just-evolving ideas about basic principles of quantum physics and complementarity." But when the two nonphysicists got lost, they interrupted for an explanation of some incomprehensible points. "It was a typical Bohr lecture such as we all have experienced so often," Courant said, "excitingly inspiring, though neither acoustically nor otherwise completely understandable." When interrupted, though, Bohr protested angrily: "Of course you cannot understand what I try to say now; this may perhaps become understandable, but only after you have heard the story as a whole and have understood the end."

At the fifth Solvay Conference in October 1927, all the founders of the quantum theory were present: Max Planck, Albert Einstein (a founder in spite of himself), Niels Bohr, Louis de Broglie, Werner Heisenberg, Erwin Schrödinger, Wolfgang Pauli, and Paul Dirac.

At session after session, Einstein sat silent. At one point he voiced a simple objection to the probability interpretation. Then he fell back into silence. He had declined an invitation to give a paper on quantum statistics at the conference.

But not all the important discussions took place on the floor of the meetings. As described by Otto Stern, the informal discussions in cafés and breakfast rooms tended to exude far more vitality:

Einstein came down to breakfast and expressed his misgivings about the new quantum theory, every time [he] had invented some beautiful experiment from which one saw that [the theory] did not work . . . Pauli and Heisenberg, who were there, did not pay much attention, "ach was, das stimmt schon, das stimmt schon" [ah, well, it will be all right, it will be all right]. Bohr, on the other hand, reflected on it with care and in the evening, at dinner, we were all together and he cleared up the matter in detail.

The debate continued at the next Solvay Conference (on magnetism), held in 1930. Einstein came prepared. He had been thinking, and he believed he had come up with an example that disproved the validity of the uncertainty principle. Imagine, he said, that you have a box. A hole in one of the walls of the box can be opened and closed by a shutter, which is controlled by a clock inside the box. You fill the box with radiation and weigh the box. Set the shutter to open just long enough for a single photon to escape. Afterward, weigh the box again. At this point, Einstein maintained, the experimenter has succeeded in finding both the photon energy and its time of passage—which is exactly what Heisenberg's uncertainty principle says cannot be done.

Bohr was stumped—shocked, in fact. With his clock-in-a-box thought experiment, Einstein had apparently found a way around the dilemma of uncertainty that Heisenberg had posed. As one observer later wrote:

During the whole evening [Bohr] was extremely unhappy, going from one to the other and trying to persuade them that it couldn't be true, that it would be the end of physics if Einstein were right, but he couldn't produce any refutation. I shall never forget the vision of the two antagonists leaving . . .: Einstein a tall majestic figure, walking quietly, with a somewhat ironical smile, and Bohr trotting near him, very excited.

By the next morning, however, the tables were turned. Bohr had an answer. He had realized a point that Einstein, ironically, had forgotten: that one of the subtle effects of Einstein's general theory of relativity was the production of the very time uncertainty the thought experiment was supposed to disprove. Bohr used Einstein's same idea of a clock in a box, but he added realistic experimental details that Einstein hadn't bothered with. Later he used a diagram to illustrate the point. The initial weighing is done by recording the position of a

Albert Einstein and Niels Bohr during one of the Solvay Conferences (probably 1930). (Photo by Paul Ehrenfest, Niels Bohr Archive, Copenhagen)

pointer attached to the box suspended from the scale, which is attached to a fixed frame. The weight measurement is uncertain (with the pointer returning to its initial position when the photon escapes). And the uncertainty of the position of the clock in the gravitational field also implies another uncertainty in the determination of time.

Bohr had showed that if you actually performed the experiment, with real equipment, the accuracy with which the energy of the photon is measured would restrict the precision with which its moment of escape can be determined. Einstein's thought experiment did not violate the uncertainty relations for energy and time, after all. Bohr provided details on all the facets of the apparatus that would be required to take the measurements in a classical physics setting. Heavy bolts hold the frame steady. The spring keeps the box mobile in the gravitational field. The weight of the box readjusts the position.

Bohr's ability to think literally had served him once again. Sometimes his literal-mindedness crept into strange contexts, however. One of Bohr's young colleagues from those days recollected times when he and his friends would talk Bohr into an evening at the movies:

> We had great preference for bad films. Sometimes we could entice Bohr to come with us to see a Western or a gangster film we had selected. His comments were always remarkable because he used to introduce some of his ideas on observations and measurements. Once, after a thoroughly stupid Tom Mix film, his verdict went about as follows: "I did not like that picture; it was too improbable. That the scoundrel runs off with the beautiful girl is logical; it always happens. That the bridge collapses under their carriage is unlikely, but I am willing to accept it. That the heroine remains suspended in mid-air over a precipice is even more unlikely, but again I accept it. I am even willing to accept that at that very moment Tom Mix is coming by on his horse. But that at that very moment there should be a fellow with a motion picture camera to film the whole business—that is more than I am willing to believe."

These were years of great discovery. Naturally, debate ensued. It is the job of science to question, and Einstein and others who questioned were doing their jobs as scientists, just as those who set forth hypothe-

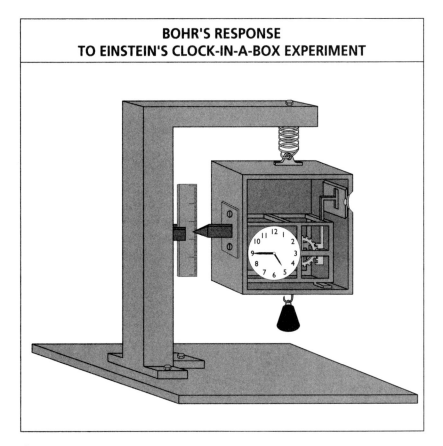

**BOHR'S RESPONSE
TO EINSTEIN'S CLOCK-IN-A-BOX EXPERIMENT**

Figure 7.

ses were doing their jobs. As the physicist Robert Oppenheimer would later write of this period in physics:

> Our understanding of atomic physics, of what we call the quantum theory of atomic systems, had its origins at the turn of the century and its great synthesis and resolutions in the nineteen-twenties. It was a heroic time. It was not the doing of any one man. It involved the collaboration of scores of scientists from many different lands, though from first to last the deeply creative and subtle and critical spirit of Niels Bohr guided, restrained, deepened and finally transmuted the enterprise. It was a period of patient work in the laboratory, of crucial experiments and daring action, of many false starts and many untenable

conjectures. It was a time of earnest correspondence and hurried conference, of debate, criticism and brilliant mathematical improvisation. For those who participated it was a time of creation. There was terror as well as exaltation in their new insight.

CHAPTER 4 NOTES

p. 57   "Of course . . . Zeeman effect!" Quoted by C. P. Snow in *The Physicists,* p. 57.

p. 59   "One does not get . . . " Born in *Zeitschrift Physic,* 1926, vol. 37, p. 863, quoted by Abraham Pais, *Subtle Is the Lord: The Science and the Life of Albert Einstein,* pp. 442–443.

p. 59   "the single . . . Newton," Lederman, *The God Particle,* p. 171.

p. 59   "Quantum mechanics is . . ." From a letter to Born, December 4, 1926; *The Born–Einstein Letters,* p. 90, quoted by Pais, *Subtle Is the Lord,* p. 443.

p. 59   "What nature demands . . ." Quoted by Pais, *Subtle Is the Lord,* p. 443.

p. 60   "The very nature . . ." Niels Bohr, quoted in Pais, *Subtle Is the Lord,* p. 444.

p. 60   "It is a bit as if . . ." Otto Frisch, *What Little I Remember,* p. 93.

p. 62   "But, but, but . . ." Quoted by Frisch, *What Little I Remember,* p. 95.

p. 62   "I think it is safe . . . like that." Quoted by Heinz R. Pagels, *The Cosmic Code: Quantum Physics as the Language of Nature,* p. 135.

p. 62   "grand period . . . the end." Richard Courant, "Fifty Years of Friendship," in *Niels Bohr: His Life and Work as Seen by His Friends and Colleagues,* S. Rozental, ed., pp. 302–303.

p. 63   "Einstein came down . . ." Quoted in Abraham Pais, *Niels Bohr's Times: In Physics, Philosophy, and Polity,* p. 318.

p. 63   "During the whole evening . . ." Léon Rosenfeld in *Proceedings of the Fourteenth Solvay Conference* (New

York: Interscience, 1968), p. 232; quoted by Pais, *Subtle Is the Lord,* pp. 446–447.

p. 65   "We had great preference . . ." H. B. G. Casimir, "Recollections from the Years 1929–1931," in Rozental, ed., p. 112.

p. 66–67   "Our understanding . . ." Robert Oppenheimer, quoted by Robert Jungk in *Brighter Than a Thousand Suns: A Personal History of the Atomic Scientists,* pp. 8–9.

5

WAVES OR PARTICLES? COMPLEMENTARITY AND THE "SPIRIT OF COPENHAGEN" 1930–1938

During the early 1930s, Bohr guided the physicists of his institute into the realm of theoretical nuclear physics. The new era began in August 1928, when George Gamow, a talented young physicist from Odessa, stopped at Copenhagen on his way back to the Soviet Union after his postdoctoral work in Göttingen. As he later told the story, he detoured to Copenhagen to meet Bohr, but had only $10 in his pocket, lodging for only one night, and no food. He urgently told Bohr's secretary that he had to see Bohr that day—he couldn't afford to stay longer. So Bohr met with Gamow and discussed with him the paper on gamma rays that Gamow had been working on in Germany. Then Bohr said, "My secretary told me you cannot stay more than one day because you have no money. Now if I organize for you a fellowship . . . would you stay for a year?" "Yes, I would," Gamow replied quickly. And so it was done, on the spot, and Gamow stayed not one but two years, from 1928–29.

During the last month of his stay, Gamow completed his book on radioactivity and atomic nuclei. It was the first book on theoretical nuclear physics ever written. So began an era of nuclear investigation at Bohr's institute.

It was a field full of major activity in the 1930s. The stage had been set by Rutherford, in England, in 1919. Continuing his experiments, he decided to try firing alpha particles through a tube of nitrogen gas. When he examined the results, he discovered something strange. In

The 1930 Copenhagen Conference. Sitting in the front row, left to right: Oskar Klein, Niels Bohr, Werner Heisenberg, Wolfgang Pauli, George Gamow, Lev Landau, and Hendrik Kramers. (Niels Bohr Institute, Courtesy AIP Emilio Sergè Visual Archives)

addition to his alpha-particle bullets (emitted naturally by radioactive radium), he found particles that had the same properties as hydrogen nuclei, even though the tube contained no hydrogen. (These particles later were called protons.) Rutherford concluded that the alpha particles had struck a few nitrogen nuclei and split the hydrogen nuclei away. While the discovery of radioactivity had shown that certain atoms could disintegrate spontaneously in nature, this was the first evidence that ordinary (nonradioactive) atoms were not indestructible. Over the next five years, with his colleague James Chadwick, Rutherford tried similar experiments skimming off protons from the nuclei of 10 different elements with alpha bullets. From these experiments they began to realize that the nucleus was not just a tiny positively charged entity; they were probing it, and if it could be probed, and chunks could be broken off in this way, it must have an internal structure—it must be composed of particles. The question was, what particles?

Perhaps, Rutherford hypothesized, there was a third, neutral, atomic particle not yet discovered. It was a reasonable idea, since protons alone could not account for the atomic weight of any atom except hydrogen. Lithium, for example, has only three protons, and yet its atomic weight is seven. The negative charge of electrons balanced the protons' positive charge. Yet the mass of electrons, the only other known particle, was negligible. Where did all the rest of the weight come from?

Irène and Frédéric Joliot-Curie, in the tradition of Irène's parents, Marie and Pierre Curie, also were actively exploring atomic structure, in Paris. In their experiments they had been bombarding beryllium with alpha particles, and they were getting a strange particle they thought was a type of penetrating radiation, like a gamma ray. Bohr wrote to Rutherford, however, that he didn't think they were right. James Chadwick, who was working with Rutherford, tried a similar series of experiments. He spent day and night at the lab for three weeks straight, finally coming up with proof of the existence of Rutherford's neutral particle.

Rutherford shot the news off to Bohr, who was delighted. In Copenhagen, he and his colleagues confirmed the results. The neutron was discovered, successfully accounting for all the excess weight on the periodic table.

Bohr now became intrigued by the dynamics of the nucleus, and in April 1932 he gave a lecture on the neutron's properties. He wrote to Heisenberg, drawing him into the quest, and Heisenberg sat down and roughed out a basic proton-neutron model of the nucleus. "The basic idea," he wrote to Bohr in June, "is to shove all difficulties of principle onto the neutron and then apply quantum mechanics to the nucleus." It was a beginning.

During these years Bohr was at the nexus of communication among colleagues all over the world, and his interpretative advice was greatly respected. He loved a paradox: Out of paradox, he believed, would come progress. And he had a talent for giving focus, meaning, and direction to lines of inquiry. In this way, he influenced the scientific process in nearly every country in the world.

By this time the study of nuclear physics began to demand ways of accelerating the bombarding particles more effectively, and in England John Cockroft and Ernest Thomas Sinton Walton invented the first particle accelerator in 1929. By 1935 Bohr had negotiated for a similar model for the institute at Copenhagen, and design studies began for its construction. Though the laboratory was small, Bohr wanted it to remain on the cutting edge of the field of nuclear physics.

Meanwhile, Bohr had been working on a new theory, featuring what he called a compound nucleus. The idea that the nucleus was a single rigid body, he said, was incorrect and misleading. Further, he maintained, nuclear reactions took place in two stages. In the first stage, during bombardment of a nucleus by a projectile, the projectile (of whatever kind—alpha particle, neutron, and so on) merges with the nucleus, forming a compound nucleus. In the second stage, this compound nucleus comes apart—in one of three ways. It may disin-

Bohr loved to talk, often far into the night, with the many young scientists who came to the institute. Here Bohr (left) delves into deep discussion with James Franck (middle) during one of the German scientist's visits in the early 1920s. (Courtesy AIP Emilio Segrè Visual Archives, Margrethe Bohr Collection)

tegrate into the original particles that made it up, unchanged by the bombardment. Or it may break up into the same particles, but excited (that is, raised to a higher energy level). Or it may form new particles (through nuclear reaction). Bohr's ideas received rapid and wide acceptance in the nuclear physics community and produced considerable impetus for further testing and extension of his model.

In December 1931, Niels Bohr was invited to become a permanent resident at the Residence of Honor on the grounds of the Carlsberg breweries, an honor bestowed for life on the citizen of Denmark considered the most prominent in the sciences or the arts. The vote by the governing committee in this case was unanimous. Niels and Margrethe moved to Carlsberg in the summer of 1932, with their six sons; their first house guests, appropriately, were Ernest and Mary Rutherford, who visited in September of that year.

Despite Bohr's stature and influence—by now he had received dozens of awards, including the Faraday medal, in addition to the Nobel Prize—his rapport with his students and colleagues remained high-spirited and open, as evidenced by the following parody of *Faust,* staged by some of his students for an annual September celebration, sometime in the early 1930s. Wolfgang Pauli played the part of Mephisto, and the Lord, played by another student, is obviously meant to represent Bohr, who was present in the audience:

BOHR (the Lord):
> Hast thou naught else to say?
> Comest thou ever with complaining?
> Is physics never to thy mind?

PAULI (Mephisto):
> Nay, 'tis all folly! Rotten, as ever, to the core!
> E'en in my days of dule it grieves me sore
> and I must ever plague these physicists the more.

B.: (*In the mixture of German and English he always used when excited*): Oh, it is dreadful! In this situation we must remember the essential failure of classical concepts . . . *muss ich sagen . . .* just a little remark . . . what do you propose to do with mass?

P.: What's that got to do with it? Mass? We must abolish it!

B.: Well, that's very, very interesting . . . but . . . but—
P.: No, shut up! Stop talking rubbish!
B.: But . . . but—

P.: I forbid you to speak!
B.: But Pauli! Pauli! We're really much more in agreement than you think! Of course, I quite agree! Only . . . Certainly, we can abolish mass. But we must uphold load. . . .
P.: What for? Why? No, no, that's wishful thinking! Why not abolish load, too?
B.: I must ask . . . I understand perfectly, of course . . . but . . . but—
P.: Silence!
B: But Pauli, you must really give me a chance to finish what I have to say! If both mass and load are abolished, what have we got left?
P.: Oh, that's quite simple! What we've got left will be the neutron!
(*Pause. Both pace to and fro.*)
B.: It's not to criticize, it's but to learn.
I take my leave now, later to return.
(*Exit.*)
P.: (*Soliloquizes*):
I like to see the old chap now and then
and take good care we don't fall out. It's jolly
decent of so grand a Lord, I must say, when
he comes himself for a nice chat with Pauli!
(*Exit.*)

By 1935 Otto Frisch had arrived in Copenhagen and was deeply impressed by the vibrant man who ran the institute. As Frisch wrote in his memoirs:

Niels Bohr was fifty then and at the height of his powers, both mentally and physically. He was heavily built, but when he thundered up the stairs two steps at a time we young ones found it hard to keep up with him. He also beat us all at table tennis.

In the evenings, the students usually went to Carlsberg for food and talk.

After dinner, we would sit around Bohr, some of us on the floor at his feet to watch him first fill his pipe and then to hear what he said. . . . Our conversation ranged from religion to genetics, from politics to modern art. I don't mean to say that Bohr was always right, but he was always thought-provoking and never trivial. How often did I cycle home through the streets of Copenhagen, intoxicated with the spirit of Platonic dialogue!

Dialogue was Bohr's strong point, and in 1935 he again had an opportunity to try his skill against Einstein in their ongoing debate over the quantum theory and complementarity. In that year Einstein lodged a counterproposal, in a paper he published jointly with Boris Podolsky and Nathan Rosen (usually referred to as the EPR paper, by the last initials of the three authors). In this volley, Einstein, Podolsky, and Rosen argued that the quantum theory was incomplete. They proposed a thought experiment (known as the EPR experiment) to show that all objects—and quantum objects in particular—possess a physical reality that is independent of whether they are observed, measured, or disturbed in any way. This is a fundamental point of difference between classical and quantum physics. If a molecule composed of two atoms is taken, EPR argues, and the two atoms are split, sending one in one direction and the other in the other direction without disturbing their spins, it can be determined by looking at one—without looking at, measuring, or disturbing the other—what the spin of the other one is.

With a classical object this would be like taking an object such as a nickel, splitting it in half through its thickness, and mailing one half to one person and the other half to yourself. When you got your half in the mail, you would know by looking only at your half whether the other person had heads or tails.

However, in the realm of the very small, if you use two measurable variables that are connected by quantum mechanics—like momentum and position—when you measure one, you disturb the other so that it cannot be measured accurately at the same time. You can know momentum or position, but not both. Those who accept quantum theory resign themselves to this inability to know. Einstein, however, refused all such renunciations. He called this disturbance of one particle while looking at another "ghostly action at a distance." While he conceded that quantum theory did explain much, he maintained it was incomplete. Bohr offered answers to EPR, but none of them satisfied Einstein, who remained unconvinced for the rest of his life.

In 1964, after both men had died, John S. Bell brought the issue to a boil again with a general calculation for an experiment modeled on EPR, but incorporating quantum theory's restriction. He found that the probability of observing matching spins of the two separated particles was different from the probability implied by EPR. Bell's

Albert Einstein in his study, c. 1953. (Photograph by Alan Richards, courtesy of the Archives, Institute for Advanced Study, Princeton, N.J.)

theorem created a great stir, seeming to offer proof for the quantum side of the argument. Perhaps, as Edward Speyer suggests in his book *Six Roads from Newton,* the ultimate solution will lie in interpreting parts of the setup that have not been challenged. Perhaps, like an audience at a magic show, we have been thrown off the track by letting our attention focus on the wrong part of the experiment.

Until recently, most physicists have accepted Bohr's complementarity—also known as the spirit of Copenhagen—as the only viable way to deal with the way things seem to work in the realm of the very small. In the 1980s and 1990s, however, some physicists have again called Bohr's ideas about complementarity into question. But science historian and physicist Philip Morrison perhaps set Bohr's contribution in the clearest light when he wrote in 1985:

> The history of particle physics is an account of splendid successes, old and new, and of shortcomings and limitations too, to be expected in any single work of our finite species. They built a city shining upon a

hill, they who half a century ago founded the quantum mechanics of particles and fields, and the design of this city is understood best through Niels Bohr.

While Bohr's work saw continued successes and challenges, for Bohr and his family in their private life, the years between 1930 and 1938 were filled with tragedy and sorrow. On Sunday morning, November 30, 1930, his mother died, after a long battle with cancer. After Ellen Bohr's death, Niels's sister, Jenny, suffered a breakdown at the loss and was placed in a mental hospital. Two and a half years later, on May 5, 1933, she also died. Her death certificate diagnosed a state of manic depression. Niels was in Pasadena, California, at the time and heard the news by cable from Harald. Niels wrote of Jenny's warm and inspiring disposition, and both brothers felt sorrow that her illness, which she had apparently fought throughout her life, had kept her from fulfilling the potential she clearly had. When Harald spoke at her funeral, he described her as "strong in spite of her weaknesses, and healthy in spite of her illness."

On September 25, 1933, Bohr's good friend Paul Ehrenfest committed suicide. He had first met Ehrenfest in 1918, through correspondence, and they met many times and visited in each other's homes in the years that followed. But Ehrenfest, who was also friends with Einstein, had become distraught in recent years over the irreconcilable differences between Bohr and Einstein. By 1927 he said he felt he had to make a choice and sided with Bohr. By 1931 deep depression had set in. In May of that year, Bohr received a letter from him in which he said, "I have completely lost contact with theoretical physics. I cannot read anything any more and feel myself incompetent to have even the most modest grasp about what makes sense in the flood of articles and books. Perhaps I cannot at all be helped any more. Still I have the illusion that you could show me the way in a few days' encounter." But by September he wrote again to say that he was too depressed even to attend conferences.

Ehrenfest wrote a letter of desperation to Bohr, Einstein, Franck, and several other of his friends in August 1933, but he never sent it. It was found after he shot his son, Wassik, who had Down's syndrome, and himself in Amsterdam on September 25. Paul Dirac later said with great regret that he should have known that Ehrenfest was contemplating suicide, when his friend remarked to him earlier that month in Copenhagen, "Maybe, a man such as I feels he has no longer the force to live." Dirac wrote to Bohr, "I now cannot help blaming myself for not doing anything." Bohr also must have felt deep regret and sorrow at this tragic loss of a friend.

Four years later Bohr lost another close friend, his great mentor and ally Ernest Rutherford. Bohr heard the news while he was attending a conference in Bologna, Italy. Rutherford died quietly on October 19, 1937, at the age of 66, from complications following an operation for a strangulated hernia. In a talk he delivered the next day in Rutherford's memory, Bohr said, "As has been said of Galileo, . . . he left science in quite a different state from that in which he found it. . . . He will be missed more, perhaps, than any scientific worker has ever been missed before." He would also be greatly missed as a friend.

But the greatest tragedy of these years for the Bohrs occurred in July 1934. Eight years earlier, Niels had purchased a yacht, the *Chita,* with three friends, including his boyhood friend Ole Chievitz. The four had spent many joyful days sailing the waters around Denmark, setting their sails to the winds, soaking up the sun, and—with Niels on board—always discussing the application of physics to the phenomena they saw around them, from the way the sails tacked in the wind to the patterns of moonlight on the water.

On their first summer outing in 1934, Niels's 17-year-old son Christian, his oldest, was on board. It all happened very fast. The seas had been rough—certainly they had sailed before in rough waters— but on this day a huge breaker suddenly rocked the *Chita* from the port side, swinging the tiller wide and sweeping young Christian overboard. A strong swimmer, he was able to keep afloat at first, and the men threw him a life preserver. But the turbulent waters were too rough. The young man could not swim to it, and he finally disappeared beneath the waves. It was all Niels's friends could do to hold him back from jumping after his son. By that time the seas were so strong that he too would have drowned if he had attempted a rescue. Finally the *Chita* docked at a Swedish port just south of Gøteborg, and from there Niels, with great sadness, made his way to Tisvilde to tell the terrible news to Margrethe and the rest of the family. The boy's body was not found until the end of August.

Other deep difficulties loomed on the horizon. In Germany a great storm was brewing. In 1933 Hitler promulgated the first of many anti-Jewish laws, stripping "non-Aryan" academics of their posts. A great number of scientists—ultimately including more than 100 physicists—began to leave Germany.

Yet in the first 40 years of the 20th century, physics had become a hotbed of excitement. From the first amazing discoveries about radioactivity and its nature, to new ideas about the nature of the atom and the recognition that most of its mass existed at its center; from the wave-particle duality to the uncertainty principle to complementarity, the surprises never seemed to cease. Strangely, the stage, it seemed, had been set for the most astounding—"fateful," it's been said—event of all, which would take place in 1938, on the eve of World War II: the splitting of uranium atoms, which emitted further particles, which in turn could split other atoms, with the possibility of a chain reaction and the release of enormous amounts of energy.

To those who were immersed in the field of physics, the news of this unheard-of result from a laboratory near Berlin simultaneously sent two shivers down the spine: one, at the recognition of an extraordinary scientific fact—that the uranium atom could be split in this way; and a second, at the comprehension of its political implications, just as Europe was entering a war with Germany.

CHAPTER 5 NOTES

p. 69   "My secretary . . ." Quoted from an interview by C. Weiner on April 25, 1968, by Abraham Pais, *Niels Bohr's Times: In Physics, Philosophy, and Polity,* p. 324.

p. 71   "The basic idea . . . nucleus." Quoted by Pais, p. 332.

p. 73–74 "Hast thou . . . (Exit.)" Reproduced by Robert Jungk in *Brighter Than a Thousand Suns: A Personal History of the Atomic Scientists,* pp. 39–40.

p. 74   "Niels Bohr was fifty . . ." Otto Frisch, *What Little I Remember,* pp. 90–91.

p. 74   "After dinner, . . . dialogue!" Frisch, p. 92.

p. 76–77   "The history of particle physics . . ." In "A Glimpse of the Other Side," in *Niels Bohr: A Centenary Volume,* A. P. French and P. J. Kennedy, eds. p. 350.

p. 77   "strong in spite of . . ." Quoted from a transcript of Harald's talk, by Pais, p. 408.

p. 77   "I have completely lost contact . . ." Quoted by Pais, p. 409.

p. 78   "Maybe, a man . . . not doing anything." Quoted by Pais, p. 410.

p. 78   "As has been said . . ." Niels Bohr in *Nature,* 140 (1937), p. 752.

6

FISSION, ESCAPE, AND THE ATOMIC BOMB 1939–1945

Of all those in the world of physics who were watching from outside Germany, Bohr was the first to act in response to the deepening Nazi crisis. Physicists in Germany began receiving notes from him, completely unsolicited, suggesting that they might want to come to Copenhagen to visit and discuss their plans. The subtext clearly was: I will offer you a haven and help you plan a future outside Germany. The flow of scientists from Germany to Denmark soon became steady, and through his influence, Bohr found new positions for them in the United States, in England, and in Sweden.

Among those Bohr helped flee from Germany was Lise Meitner, a physicist who had been working at the Kaiser Wilhelm Institute in Berlin with the respected German chemist Otto Hahn. Meitner was also the aunt of Otto Frisch, who had escaped German rule much earlier and had been working at Bohr's institute for several years. She left in 1938, and Bohr found her a position in Stockholm at the new physics laboratory the university was building there.

For 30 years she had worked side by side with Hahn. He typically did the experimental work, she the theoretical interpretation. Now, at 60, her laboratory and her partner remained in Berlin, and she had almost no equipment to work with at the not-yet-built laboratory at the new Physical Institute in Stockholm, where she had been lucky even to find a position. So when Christmas holidays approached, she gladly accepted the invitation of friends to visit them on the northeast coast of Sweden in Kungälv, a little resort town where they lived.

There she had arranged to meet her nephew, Otto Frisch, who was surprised to find his aunt completely absorbed in a letter from Hahn.

The two set out in the snow, Frisch on skis, Meitner on foot, insisting that she could keep up. The two puzzled over the news, which had come across from Germany in very garbled and excited letters. Meitner pulled scraps of paper out of her pocket to make calculations.

Hahn and his assistant, Fritz Strassmann, had been trying to solve a mystery by bombarding small quantities of uranium with neutrons. This process typically produced only a few thousand atoms of "daughter substances," new substances having a different atomic makeup from the parent, which in this case was uranium. Then the challenge was to identify the new substances and explain why they were produced.

But Hahn wrote Meitner because he could not understand his results. "The fact is," he wrote, "there's something so strange about the 'radium isotopes' [produced by the bombardment] that for the time being we are mentioning it only to you. . . . Our radium isotopes act like *barium*."

Hahn and Strassman expected to split little pieces off the nucleus, somewhat as Rutherford had, to produce substances that were nearby on the periodic table, such as radium, with an atomic number of 88, compared with uranium's 92. Barium, however, has an atomic number of 56. As Frisch and Meitner walked, they realized that Hahn and Strassmann had done the undoable. They had split the heavy uranium atom asunder.

Both Meitner and Frisch also realized the implications of this capability in the hands of German Nazis. The relatively enormous quantity of energy released by splitting uranium atoms could have great destructive power. The conclusions they had reached, making their calculations as they sat in the snow, were also very exciting news for the scientific community. This was news that must reach the right people.

Frisch hurried back to Denmark. Arriving just before Bohr's departure by steamer for a meeting in the United States, he set the evidence before his mentor.

"Oh what idiots we all have been! Oh but this is wonderful! This is just as it must be!" exclaimed Bohr, striking his hand to his head. Excited, he encouraged his protégé to publish a paper with Meitner on their interpretation of the Hahn-Strassman results as soon as possible. Then he embarked for the United States. En route, he mentioned the exciting news to a colleague, and at a meeting the news leaked out—before Meitner and Frisch's paper was published, as it turned out (a slipup that Bohr always regretted). Literally overnight, physicists and chemists in universities all over the United States began testing the premise and found it was true. The atom had been split! The worlds of physics and chemistry had been flipped upside-down.

Bohr, of course, was devastated that he had leaked a discovery by other scientists before they had a chance to publish, and rushed to write a note for publication in explanation.

The ultimate result of Hahn and Strassmann's experimental results, which would have been published (and were) in any case, was that the United States and Great Britain decided to build an atomic bomb, out of fear of this knowledge in German hands.

In 1940 the German threat moved closer for Bohr, when the German armies invaded his homeland. He and Margrethe decided to stay, however, and in the following three years, he became prominent in efforts to get those in danger out of Denmark, helping them find boats to cross the sea to safety.

Another concern of Bohr's, of course, was that assets and equipment should be prevented from falling into German hands, if possible. Bohr devised a means to protect two Nobel medals—Max Planck's and Max Laue's—that had been given to him for safekeeping. To prevent the Nazis from recognizing the value of the medals, he dissolved them in acid and hid the acid-filled vials. After the war he precipitated the gold out of the acid and recast the medals in their original form for return to their owners. (Bohr had donated his own medal to aid the Finnish war effort.)

Bohr stayed in Denmark until 1943, when it became clear that the Germans intended to imprison him. He and his family narrowly escaped in a fishing boat, and once he reached safety in Sweden he continued to help many others escape. There the decision was made to fly him with his son Aage, then 21, to London, where they could help with the British war effort. Bundled into the cramped bomb bay of an unarmed Mosquito British bomber, Bohr became preoccupied during the flight and pushed off his helmet. Bohr didn't hear the pilot's warning that they would be flying to high altitude, and by the time the plane arrived in London, he was near death from lack of oxygen.

Niels and Aage journeyed to the United States in late 1943 as representatives of the British. There security authorities were anxious to keep the bomb-building Manhattan Project as secret as possible—the psychological power of an unexpected attack potentially was an even mightier weapon than the bomb itself. They worried that the Bohrs' presence in the United States might be noticed and questioned and that some comment in a newspaper might tip spies off to the existence of the project. So, on arrival, officials issued false identification papers to Niels and Aage, disguising Niels as "Mr. Nicholas Baker," and his son as "James Baker." (Niels, of course, forgot that his real name was inscribed on his suitcase!) Throughout their travels they were accompanied by an armed bodyguard, complete with signed receipts for their well-being required at every changing of the guard.

Under guard, the two proceeded circuitously to a pine-dotted mesa called Los Alamos in northern New Mexico. There the United States government had gathered the best physicists in the country to work on a secret project—the development of the atomic bomb. Aage, whom everyone called "Jim," became a junior scientific officer, and Niels ("Uncle Nick") became consultant to the directorate, headed by Robert Oppenheimer, a tall, lean man who also smoked a pipe and whose intelligence and energy radiated in Los Alamos as Niels's did in Copenhagen.

"Bohr at Los Alamos was marvelous," Oppenheimer would later say in a lecture after the war. "He took a very lively technical interest . . . But his real function, I think for almost all of us, was not the technical one. He made the enterprise which looked so macabre seem hopeful."

In later reflections Oppenheimer defined Bohr's vision as a kind of complementarity of the bomb—a weapon of great destruction that, unused, could be the key to a different world. He brought this message to the scientists at Los Alamos. Victor Weisskopf, a young physicist who had fled from Austria and worked at Los Alamos, would later describe Bohr's contribution in these terms:

> In Los Alamos we were working on something which is perhaps the most questionable, the most problematic thing a scientist can be faced with. At that time physics, our beloved science, was pushed into the most cruel part of reality and we had to live it through. We were, most of us at least, young and somewhat inexperienced in human affairs, I

would say. But suddenly in the midst of it, Bohr appeared in Los
Alamos.

It was the first time we became aware of the sense in all these terrible
things, because Bohr right away participated not only in the work, but
in our discussions. Every great and deep difficulty bears in itself its own
solution. . . This we learned from him.

By mid-February 1944 Bohr was back in Washington, where he
contacted a friend who was a justice of the U.S. Supreme Court, Felix
Frankfurter. Through carefully circumspect communication, they
established that they both were aware of Los Alamos and the Manhattan Project. Through Frankfurter, Bohr communicated with President
Roosevelt about his concerns regarding the postwar future. From the
communications from Roosevelt, Frankfurter and Bohr decided that
the president wanted Bohr to act as go-between to Winston Churchill,
prime minister of Britain, to express the president's concern and his
interest in including the Soviets among those informed about the

Robert Oppenheimer at Los Alamos. (Los Alamos National Laboratory)

Always athletic, even in later years, former soccer player Bohr skied the snowy slopes surrounding Los Alamos during his work there. (Los Alamos National Laboratory)

bomb project. The idea was that if the U.S. and Britain confided in the Soviets early, postwar arms control might result, whereas failure to include the Soviets would almost certainly lead to an arms race.

Niels and Aage went to London, apparently with a mandate to talk with the prime minister, and on May 16 Niels set out for 10 Downing Street. Aage, who was Niels's complete confidante and acted as his secretary during this period, remembered what happened in this way:

> We came to London full of hopes and expectations. It was, of course, a rather novel situation that a scientist should thus try to intervene in world politics, but it was hoped that Churchill, who possessed such

imagination and who had often shown such great vision, would be inspired by the new prospects.

The meeting went very badly. Churchill was in a bad mood, suspected his aides of subterfuge, and barely gave Bohr an opportunity to talk. When Bohr asked if he could write to Churchill, the prime minister replied, "It will be an honour for me to receive a letter from you, but not about politics." According to some sources, Churchill became so angry over the incident that he nearly had Bohr arrested.

Discouraged but unstoppable, Bohr did write to Churchill and to Roosevelt, several times, and also returned to the United States to communicate again with Roosevelt through his contact, Frankfurter. But Churchill failed to recognize that the bomb would not remain a secret forever, that if the British and Americans could build an atomic bomb, so could other nations, and that, as Bohr would put it succinctly in later life, "We are in a completely new situation that cannot be resolved by war."

First test of the atomic bomb, Almagordo, New Mexico, July 1945. (Los Alamos National Laboratory)

CHAPTER 6 NOTES

p. 82   "The fact is . . ." Richard Rhodes, *The Making of the Atomic Bomb,* p. 251.

p. 82   "Oh what idiots . . ." Quoted in Otto Frisch, *What Little I Remember,* p. 116.

p. 84   "Bohr at Los Alamos . . ." Robert Oppenheimer, quoted by Rhodes, p. 524.

p. 84–85   "In Los Alamos . . ." Victor Weisskopf, quoted by Rhodes, pp. 524–525.

p. 86–87   "We came to London . . ." Aage Bohr, quoted by Rhodes, p. 529.

p. 87   "It will be an honour . . ." Aage Bohr, "The War Years and the Atomic Weapons" in *Niels Bohr: His Life and His Work as Seen by His Friends and Colleagues,* S. Rozental, ed., p. 204.

p. 87   "We are in a completely . . ." Quoted by Rhodes, p. 532.

7

THE FINAL YEARS 1945-1962

Germany surrendered in May 1945, and for Europe the war was finally over. But war continued to rage in the Pacific, and the development of the atomic bomb was coming to a climax in Los Alamos. On July 16, 1945, the first detonation of a nuclear device occurred in a test explosion at nearby Trinity. On August 6 Hiroshima was destroyed by an atomic bomb at 8:15 AM, Japanese time. On August 9 a second bomb was dropped, this time striking Nagasaki. It was 11:02 AM, Japanese time.

Two days later an article written by Niels Bohr with Aage's help appeared in the London *Times*. Its title was "Science and Civilization," and it was Bohr's first public plea for a world where open exchange of information could prevent such horrors in the future. He wrote:

> Civilization is presented with a challenge more serious perhaps than ever before. . . . We have reached the stage where the degree of security offered to the citizens of a nation by collective defense measures is entirely insufficient. . . . No control can be effective without free access to full scientific information and the granting of the opportunity of international supervision of all undertakings which, unless regulated, might become a source of disaster. . . . The contribution which an agreement about this vital matter would make . . . can hardly be exaggerated.

Niels and Margrethe Bohr returned home to Copenhagen at last on August 25, where they resumed their life at Carlsberg House and Tisvilde. Bohr, as always, threw himself headlong into his work and enjoyed the company of his grandchildren, for whom he loved to buy toys and gadgets. In some ways his life went on as before; in some ways it was greatly changed.

By 1945 Niels Bohr had become an eminent public figure, not only in Denmark, but worldwide. The king and queen of Denmark came to Carlsberg House for dinner, as did cabinet ministers and ambassadors. A lengthy procession of high dignitaries and heads of state paid visits to Bohr in Copenhagen during the years following the war—including Queen Elizabeth II and Prince Philip of England, the queen of Siam, the crown prince of Japan, the prime ministers of India and Israel, and even Winston Churchill, the man who almost had him thrown in prison during the war. As always, he also entertained a steady parade of physicists, both senior and junior, and spent long hours talking to them and listening to their ideas.

At 60, Bohr's energy had not flagged—he still took the steps at the institute two at a time. He continued his involvement with research issues as well as the continuing growth of the institute. But also, in the years following the war, he became more and more involved in an issue that had come to trouble him greatly ever since the specter of the atomic bomb had become visible.

Even before the war drew to a close, as the Allies gained the upper hand, Bohr had constantly looked ahead, recognizing the future of distrust, fear, and possible destruction that the world's nations would face after the war. Although the Allies had cooperated well during the war, he knew that the spirit of cooperation would not be maintained once the current threat was past. Too many fundamental differences in economic and political organization existed among them. He had seen firsthand how difficult it was even to maintain scientific contacts with colleagues within the Soviet Union, as that vast Communist country grew more and more isolated. And he could foretell that finding a basis for cooperation between East and West would be a monumental, if not impossible, task.

As Aage Bohr later explained:

> It was against this background that my father saw the prospects of the atom bomb, and he realized immediately how decisively this new development would affect the world. He perceived that soon the world might be faced with an arms race that would threaten the continued existence of civilization itself. However, it was characteristic of my father's whole attitude that he also saw immediately that these very prospects offered new possibilities of giving political developments a more favourable turn. Indeed, everyone would have to realize that the world was changed for better or for worse, that now a comprehensive

and genuine co-operation was necessary to avoid living under the most ominous threats. Here was a vital task for united effort.

During the coming decade, Bohr made use of his many contacts, working tirelessly for his vision of a better world—a world where the threat of atomic weapons was defused by a policy of worldwide openness. He continued to make overtures to governments. In 1948 he met with George Marshall, the secretary of state of the United States. Marshall listened with interest, but, in the end, the U.S. government balked at backing Bohr's policy.

By August 1949 the Soviet Union had exploded its first test bomb. The United States made a public announcement in 1950 that it would pursue development of the hydrogen bomb. The arms race Bohr had feared had begun.

But Bohr would not give up. He decided to make a direct worldwide appeal, no longer in private conferences with heads of state but in an open letter to the United Nations. In February 1950 he arrived once again for a short stay at the Institute for Advanced Study in Princeton, where he enlisted the help of his former student and future biographer, Abraham Pais. Many discussions ensued. Bohr returned to

Niels Bohr and his son Aage. (Courtesy of the Archives, Institute for Advanced Study, Princeton, N.J.)

Bohr at work in his study at the Carlsberg House of Honor. (Niels Bohr
Institute, courtesy AIP Emilio Segrè Visual Archives)

Copenhagen in May, where he and Harald hammered out the text of
the letter in secret. They sent the letter to Bohr's son Aage, who at the
time was at Columbia University in New York. At 10:00 A.M. on June
9, a moment timed to coincide with a press conference Bohr called in
Copenhagen, Aage Bohr delivered the letter to the office of the
secretary general of the United Nations, Trygve Lie.

The following excerpt contains the major points of the letter:

> The situation calls for the most unprejudiced attitude towards all questions of international relations. Indeed, proper appreciation of the duties and responsibilities implied in world citizenship is in our time more necessary than ever before. On the one hand, the progress of science and technology has tied the fate of all nations inseparably together, on the other hand, it is on a most different cultural background that vigorous endeavours for national self-assertion and social development are being made in the various parts of our globe. An open world where each nation can assert itself solely by the extent to which it can contribute to the common culture and is able to help others with experience and resources must be the goal to be put above everything else . . . The arguments presented suggest that every initiative from any side towards the removal of obstacles for free mutual information and intercourse would be of the greatest importance in breaking the present deadlock and encouraging others to take steps in the same direction.

Bohr opened his press conference in Copenhagen and read the contents of his letter before the assembled members of the press. Having said everything he meant to say in the text of some 5,500 words, he refused, for the most part, to accept questions and concluded his news conference.

The reaction was disappointing. Secretary General Lie acknowledged receiving the letter, assuring Bohr that he would give it most careful consideration, but the international body never made plans to discuss or debate its premises. Few outside the Scandinavian countries paid any attention to the news conference. Not until Bohr distributed thousands of copies (printed at his own expense) to the foreign news media, heads of state, colleagues, and friends did some feedback begin to drift in. But even then the response was primarily neutral or negative.

Two of the more positive reactions came from the *Washington Post* and the *New York Herald Tribune.* The first cautioned somewhat lamely against dismissing Bohr's ideas without some consideration. The New York paper maintained, only a little more supportively, "The open world is an idea of such simplicity, such power and such potential effect that it is not lightly to be dismissed."

The times were not right for the kind of trust and mutual cooperation that Bohr envisioned, and would not be for another 30 years. Instead, June 24 of 1954 marked the beginning of the Korean War, and the United States exploded its first hydrogen (thermonuclear) bomb in the Pacific before the year was out.

Bohr never gave up on his idea, though. During the last 20 years of his life, he focused primarily on the problems as well as the prospects inherent in the release of atomic energy. And he always continued to maintain that openness—free access to information and free exchange of ideas—was the only course that would be in the mutual interest of all nations.

A fine footnote to this period of Bohr's life came about in 1985 long after his death, when Mikhail Gorbachev came to power in the Soviet Union, promoting what he called glasnost, a new spirit of openness between his country and the West. That fall, all Denmark celebrated the centenary of Bohr's birth, and among the several special events that took place in Copenhagen, a three-way TV hookup was arranged, linking speakers in Copenhagen, Moscow, and Cambridge, Massachussetts to discuss the challenge of nuclear armaments. Afterward, a book of essays was published on the subject, "dedicated to Niels Bohr and his appeal for an open world."

Bohr had appealed for political action to turn atomic energy to human benefit, with no results; thereafter he turned to the scientific community. Physics, he argued, could lay the groundwork of cooperation between nations. During the 1950s he became involved in several international projects that have greatly advanced the study of physics in Europe. By this time physics had entered a new phase, with particle physics replacing nuclear physics as the cutting edge of the field. Particle physics, sometimes referred to as "high-energy" physics, took another step deeper into the mysteries of the atom. Theoretical and experimental investigations of subatomic particles required powerful (high-energy) machines, far bigger and more expensive than the accelerators of the 1930s. Clearly, if European nations wanted to recover from the "brain drain" caused by the departure of thousands of German and Italian scientists during the war, cooperation was necessary.

In part as a result of discussions at a conference in Copenhagen called by Bohr in 1951, the idea developed for a new, international center for theoretical and experimental physics. By 1952 representatives from 14 European countries met in Copenhagen to lay the groundwork for the *Conseil Européen pour la Recherche Nucléaire* (European Council for Nuclear Research), known as CERN. The cooperation, of course, did not go completely smoothly. Debates ensued about what kind of equipment was needed and where it should be located. Bohr and others had hoped that CERN might locate in Copenhagen, and for a time the theoretical branch was located there. But in the end, Geneva became the chosen site, and by 1957 a synchrocyclotron particle accelerator operating at 0.6 giga electron volts (0.6 GeV, or 1 billion electron volts) was in operation. By 1959 the proton synchrotron particle accelerator began operating at 28 GeV.

When it became clear that CERN would not stay in Copenhagen, Bohr helped organize a theoretical physics consortium, not in competition with CERN, formed jointly by an alliance of Scandinavian countries, including Denmark, Norway, and Sweden. Called Nordita (for *Nordisk Institut for Theoretisk Atomfysik*), the new institute would be centered in Copenhagen. Finland joined the group in 1955. Nordita began operations on September 1, 1957, much to Bohr's pleasure. In 1964, after Bohr's death, Nordita moved into quarters next to the Institute for Theoretical Physics. (Bohr's institute was renamed the Niels Bohr Institute the following year.) With expanded research programs in both institutes—including condensed matter physics, particle physics, and astrophysics—today the two institutes work closely together. Nordita has thrived in the more than 35 years since its foundation, having become a center for research, colloquia, and conferences that attract participants from all over the world.

A third opportunity came to light in December 1953, when President Dwight D. Eisenhower made a proposal before the United Nations, calling for an end to the atomic arms race and establishment of international cooperation to promote peaceful uses for atomic energy. While this proposal sounds like more than it was (the arms race did not come to an end), Eisenhower was implying that uranium sources might be made available to other countries. Denmark was among the first to express interest. The Danish Academy of Technical Sciences was instrumental in getting the ball rolling. A committee took form, chaired by Bohr, and in 1954 Bohr visited with the prime minister of Denmark to explore the possibility with him. In March 1955 the government formed a Preparatory Atomic Energy Commission, with Bohr at its head. Beginning in April 1955, Bohr began working on arrangements with the Danish government for establishing agreements with the United States and the United Kingdom for technical consultation on the building and construction of reactors.

Meanwhile, he began searching for a suitable site for the reactors and research facility, in many cases personally checking out possible islands and peninsulas. Finally he asked a group of commission members to take a look at a couple of sites he thought had potential. The spot he was considering was located near the medieval city of Roskilde, about 20 miles west of Copenhagen. The first site, which he thought was the less likely of the two, was an island, called Bolund, that at low tide nearly formed a peninsula. One of the commissioners told of the excursion in wonder:

> We went down to the beach. From there it seemed evident that it did not suit the purpose. Bohr was not content with guesses, however. We should have a look. Off with shoes and socks, up with pants, out into the water, Bohr ahead, over sharp stones, then up the slope to Bolund's top—our guess was correct, we could in good conscience drop Bolund and concentrate on other possibilities—in the meantime we had learned something about Bohr's working style.

By the following year Denmark was ready to set up a permanent Atomic Energy Commission, and Bohr was asked to accept the chairmanship, which he gladly did. He was 70 years old. A rocky peninsula called Risø, which jutted out into the fjord near the rejected island of Bolund, would be the site, and by the fall of 1956, construction on the research laboratories had begun. Contracts were signed to acquire three reactors. By the summer of 1957, laboratories for physics, chemistry, reactor development, and electronics were in use. A machine shop, meteorology station, and agricultural station were also in operation. A formal inauguration ceremony took place on June 6, 1958, with the king and queen of Denmark and many other Danish and foreign officials in attendance. By 1960 all three reactors were in full operation.

Under Bohr's direction, facilities at the Institute for Theoretical Physics were growing as well. A new five-story building was under construction, and an underground structure to house the revamped cyclotron was built. In addition, the institute scientists felt that a larger accelerator was needed—but there was no place left to build in the area surrounding the institute. The Atomic Energy Commission came to the rescue with an offer of land near the Risø center. Plans for building a new laboratory began at once, with an accelerator that could produce protons at 10 to 12 MeV (million electron volts)—only a fraction the size of the accelerator at the new CERN facility, but nonetheless an exciting prospect for physics research in Denmark. It first went into operation in 1961. The new laboratory is now called the Niels Bohr Institute, like the institute in Copenhagen, and it has become a vital nexus of international research.

Although Bohr's well-considered, even noble, plea for openness among nations went virtually unheeded in the political arena, he had clearly succeeded in marshaling international cooperation among scientists and in developing peaceful uses for the release of atomic energy. On October 24, 1957 the first Atoms for Peace Award—a gold medal and a check for $75,000—was presented to him at the Great Hall of the National Academy of Sciences in Washington, D.C. At that time President Eisenhower called Bohr "a great man whose mind has explored the mysteries of the inner structure of atoms, and whose spirit has reached into the very heart of man."

Between 1945 and 1961 Bohr delivered more than 20 lectures all over the world. In the process, he also received more than 30 honorary doctorates from such prestigious institutions as Princeton and Columbia universities. During the 1950s Bohr's travels were extensive, including visits to Iceland in 1951, to Israel in 1953, to Yugoslavia in 1956, to Greenland in 1957, and again to Yugoslavia and Israel in 1958.

Bohr also made numerous visits to the United States, usually spending time in residence at the Institute for Advanced Study in Princeton, where he had a standing membership. During his visits of February to June 1948, February to May 1950, and September to December 1954, he often camped in Einstein's office (Einstein preferred the office anteroom or the study at his home) and held forth in his continuing discussion with his old friend, always trying to swing him over to his own way of thinking about complementarity and quantum mechanics. He never succeeded.

Einstein's death in April 1955 hit Bohr hard. This was a friend for whom he had the deepest respect, admiration, and love. His inability to reconcile their intellectual differences gnawed at him constantly, and Einstein was on his mind for the rest of his life. In a memorial, Bohr wrote, "With the death of Albert Einstein, a life in the service of science and humanity which was as fruitful as any in the whole history of our culture has come to an end . . . He gave us a world picture with a unity and harmony surpassing the boldest dreams of the past." But for Bohr personally, much more than that, a kindred mind, a sort of twin light had gone out. "To the whole of mankind Albert Einstein's death is a great loss and to those of us who had the good fortune to enjoy his warm friendship it is a grief that we shall nevermore be able to see his gentle smile and listen to him."

Bohr suffered two other great personal losses during the 1950s. In 1951 his brother Harald died of cancer at the age of 63. Next to his wife, Harald had always been Niels's closest friend and greatest ally. The loss was huge. Seven years later, in 1958, Wolfgang Pauli died suddenly in Zurich, Switzerland. Of all the younger physicists Bohr had worked with in the 1920s and 1930s, Pauli had always understood him most intuitively. Niels and Margrethe traveled to Switzerland for the memorial services.

As his 75th birthday approached, Bohr traveled to India in January 1960, where he attended the Indian Science Congress in Bombay. There he delivered two lectures, on human knowledge and atoms, and on the principles of quantum physics. He took time to explore Calcutta, Madras, Agra, and Delhi, where he visited with the prime minister, Jawaharlal Nehru. He and Margrethe also visited the carved caves of Elephanta in Bombay Harbor, known for their eighth-century sculptured figures, as well as the caves of Ellora in central India and Ajanta in east-central India, known for their exquisite ancient sculptures, carvings, and paintings, some dating back as far as the third century B.C.

The following month he journeyed to Geneva for the inauguration of CERN, and in May of 1961 he visited the Soviet Union for the second time, where he stopped at institutions in Moscow and Dubna and spent three days in Georgia.

But by the fall of 1961, Niels Bohr was finally beginning to slow down. At the 12th Solvay conference in Brussels that October—the 50th anniversary of the first of these meetings—he presented a talk entitled "The Solvay Meetings and the Development of Quantum Physics." He asked permission to deliver the address from his seat.

Still, in June 1962 Bohr made another short trip to the United States to accept an honorary doctorate from the Rockefeller Institute in New York, and later that month he drove with Margrethe by car from

Copenhagen to Cologne, Germany, where he delivered a talk entitled "Light and Life Revisited" at the opening of the Institute of Genetics. Throughout his career, Bohr had at various times tried to apply complementarity and a mechanistic approach to biology, an effort that was rarely well received. On this occasion, he stated, in a typical effort to reset humanity's sights on the search for truth: "In the last resort, it is a matter of how one makes headway in biology. . . . Life will always be a wonder, but what changes is the balance between the feeling of wonder and the courage to try to understand."

From Cologne, the Bohrs continued on to the Bavarian city of Lindau, located on Lake Constance, where Nobel laureates traditionally gathered annually. While there Niels suddenly became ill. He was hospitalized with a diagnosis of minor cerebral hemorrhage and then was hurried back to receive medical attention in Denmark. He spent the following month recuperating with Margrethe at Tisvilde, where the two celebrated their golden wedding anniversary on August 1. It was a joyful party, attended by all the children and grandchildren—in whom Bohr always delighted enormously.

On September 30 Margrethe and Niels left for Italy, where they spent a happy vacation together in the resort town of Amalfi, overlooking the Gulf of Salerno on the southwestern coast. The sunlight sparkled on the azure waters, and the air smelled of fresh lemons and sea breezes, as colorful fishing boats bobbed in the little port. There they relaxed and celebrated Niels's 77th birthday, returning to Copenhagen on October 27.

With his doctor's blessing, Bohr resumed his regular work schedule, chairing a meeting of the Danish Royal Academy of Sciences and Literature on Friday, November 16, 1962, as he had for many years. The following day he kept an appointment to tape an interview on the history of quantum physics. His voice sounded tired.

November 18, 1962 was a Sunday, and after lunching with Margrethe and a few friends, Bohr went upstairs to take a nap. A few moments later his wife heard him call out "Margrethe," and she rushed up the stairs to his side. Niels had collapsed, unconscious, next to his bed. The diagnosis was heart failure.

Niels Bohr's remains were cremated, and burial took place at Assistens Kirkegaard, where his ashes were placed near those he loved best—his brother Harald, his son Christian, and his parents. Twenty-two years later his wife, Margrethe, was buried at his side.

The public reaction to Bohr's death was a great outpouring of admiration, respect, and affection. Physicists and friends from all over the world sent their condolences. In an unprecedented tribute to a nonmember, the *Folketing* (the Danish parliament) stood to honor him as their president talked of Denmark's sorrow. The prime minister of Israel, the king of Sweden, the chancellor of Germany, and President Kennedy of the United States all sent messages. Newspapers and magazines bore tributes such as "With the passing of Niels Bohr the world has lost not only one of the great scientists of this century but also one of the intellectual giants of all time" (the *New York Times*). And "For physicists throughout the world, Niels Bohr, the gentle genius of Denmark, has long embodied the heroic image of a scientist" (the *Christian Science Monitor*).

But perhaps the greatest representation of Niels Bohr's untiring effort to pursue knowledge remained untouched on his study blackboard—a diagram drawn the night before his death in one last effort to overcome the arguments of his longtime friend and intellectual sparring partner, Albert Einstein.

Bohr's bulldog tenacity in the long-standing debate, of course, was not one-sided. A month before Einstein's death, Einstein was also still writing about his disagreements with the Copenhagen interpretation and Bohr's arguments.

And the debate isn't over yet. Today a few physicists still have problems with what they call the "Copenhagen orthodoxy" and have begun to veer away from using Bohr's complementarity to explain the physical characteristics of the atom. If they are on the right track, Bohr would be the first to rejoice. As Otto Frisch once wrote about him, "He never hesitated for a moment to admit that he had been in error; to him it merely meant that he now understood things better, and what could have made him happier?"

Albert Einstein once said of Bohr, "He utters his opinions like one perpetually groping and never like one who believes to be in possession of definite truth." For a scientist, this was the ultimate compliment, from one of the greatest scientists of all time. It was this quality, above all, that

Bohr's last blackboard drawing. (Niels Bohr Library, Margrethe Bohr Collection).

made Niels Bohr the remarkable scientist that he was and a supreme model for anyone who is truly committed to the pursuit of truth.

CHAPTER 7 NOTES

p. 89   "Civilization . . . exaggerated." Quoted in Abraham Pais, *Niels Bohr's Times*, p. 504.

p. 90–91   "It was against this background . . ." A. Bohr, "The War Years and the Atomic Weapons," in *Niels Bohr: His Life and Work as Seen by His Friends and Colleagues*, S. Rozental, ed., p. 199.

p. 93   "The situation calls for . . ." N. Bohr, "Open Letter to the United Nations," dated 9 June 1950, reprinted in Rozental, ed., p. 350.

p. 93   "The open world . . ." *New York Herald Tribune*, June 14, 1950.

p. 96   "We went down to the beach . . ." Quoted in Pais, p. 525.

p. 97   "a great man . . ." Quoted in Pais, p. 2.

p. 97–98   "With the death of Albert Einstein . . . listen to him." In *Scientific American*, 192 (June 1955), p. 31.

p. 99   "In the last resort . . ." Quoted by Pais, p. 444.

p. 100   "He never hesitated to admit . . ." Otto Frisch, "The Interest in Focusing on the Atomic Nucleus," in Rozental, ed., p. 140.

p. 100   "He utters his opinions . . ." Quoted in Abraham Pais, *Subtle Is the Lord: The Science and Life of Albert Einstein*, p. 417.

EPILOGUE
THE LEGACY OF NIELS BOHR

Bohr at the Institute for Advanced Study. (Courtesy of the Archives, Institute for Advanced Study, Princeton, N.J.)

When Niels Bohr died on November 18, 1962, a great light went out, the illuminations of a brilliant and humane mind were stilled, and, as many sadly put it at the time, the last of the great giants had gone.

Even more than Einstein, Niels Bohr had created new ways of looking at the world in the first half of the 20th century. With the Bohr atom, science began a journey that continues to amaze, disturb, and

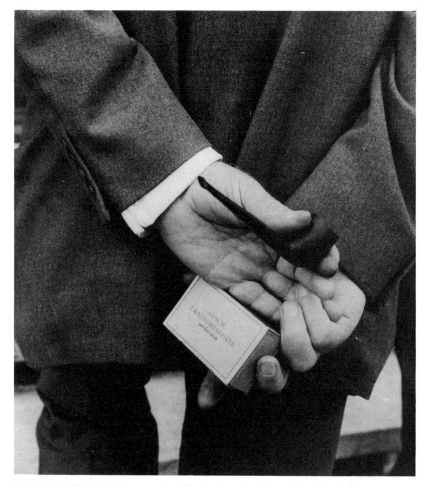

Niels Bohr with his ever-ready pipe in hand. (Niels Bohr Archive)

enlighten today. Under his stewardship the quantum revolution brought to light some of the finest minds of 20th-century physics, a truly astonishing group of brilliant and innovative scientists, who in turn sowed the ideas that have so profoundly changed our world and the way that we attempt to understand it.

The gradual revelation of the quantum nature of the world has resulted in such practical applications as lasers, MRI (magnetic resonance imaging) machines, and semiconductors (the basic materials for transistors and chips that today make possible electronic marvels ranging from miniature radios and television sets to digital watches and personal computers). The much talked-about "Information Superhighway" of the not-so-distant future will have at its core an array of technological marvels based on our understanding and application of quantum theory.

Perhaps even more important, quantum theory and quantum mechanics have changed our understanding of nature at a very basic and profound level. That this new understanding is disturbing to many even today is tribute to its fundamentally revolutionary nature.

Will we see yet another revolution in physics, or with the quantum revolution have we come to the end of the story? Today some scientists are still divided in their acceptance of the "Spirit of Copenhagen." Is the quantum theory complete, or will tomorrow bring a new and perhaps more startling revolution?

Whatever the answer to this question, there can be no doubt that in the annals of scientific history, Niels Bohr will continue to stand as a bright beacon. As both scientist and human being he was a man of insatiable curiosity and tremendous perseverance, a man of unusual intellectual talents and love of knowledge. It was a love that he shared with all around him, with a uniquely infectious enthusiasm. Almost alone in the pantheon of great scientific geniuses such as Galileo, Newton, and Einstein, Niels Bohr came most alive, as a thinker and a human being, in the company of others. Few who came in contact with Bohr were not inspired to put forth their best and develop their deepest thoughts. Out of this exciting interaction, the give and take, inventiveness and criticism, came some of the most profound ideas of our time. And out of it too came a gallery of great thinkers and scientists. Whether students, friends, or coworkers (and often many were all three at the same time), all who knew and worked with Bohr carried with them forever afterward a part of the Bohr legacy.

GLOSSARY

alpha particle (alpha rays): one of three types of radiation (the others are beta particles and gamma particles) discovered in early studies of radioactivity around 1900. Rutherford showed that alpha particles are helium nuclei, which are made up of two protons and two neutrons tightly bound together.

atom: the smallest chemical unit of an element, consisting of a dense, positively charged nucleus surrounded by negatively charged electrons. The Greek thinker Leucippus and his student Democritus originally conceived of the idea of the atom in the 5th century B.C. as the smallest particle into which matter could be divided. (The word "atom" comes from the Greek word *atomos,* which means "indivisible.") But in the 1890s and early 20th century, scientists discovered that the atom is made up of even smaller particles, most of which are very strongly bound together.

atomic mass (also known as atomic weight): the relative average mass in atomic mass units (or amu, equal to 1/12 the weight of the isotope carbon-12) of the masses of all the isotopes found in a natural sample of an element.

atomic number: the number of protons in the nucleus of an atom.

atomic spectrum: the characteristic pattern produced by light emitted by any element when it is heated.

beta particle: a high-speed electron (or its positive cousin, the positron) emitted during radioactive decay.

complementarity: the existence of different aspects of the description of a physical system, seemingly incompatible but both needed for a complete description of the system. In particular, the wave-particle duality (the need to describe some systems as composed of both waves and particles).

cyclotron: a device used to accelerate atomic particles around and around in a circle until they reach immense energies.

electron: a very light, negatively charged subatomic particle that orbits the nucleus of an atom.

fission: a nuclear reaction in which an atomic nucleus splits into fragments. Fission can generate enormous amounts of energy.

isotope: one of two or more of any element's atoms with the same number of protons in the nuclei, but a different number of neutrons. Isotopes have the same chemistry but different nuclear physics.

neutron: an uncharged, or neutral, particle found in the nucleus of every atom except hydrogen. It has nearly the same weight as the proton.

photon: a small particle, or "packet," of electromagnetic energy, having no mass and no electric charge; electromagnetic force when it behaves as a particle, rather than as a wave. The photon's existence was first proposed by Albert Einstein to explain the behavior of light in photoelectric experiments.

Planck's constant (h): the quantity that underlies all of quantum physics. It has the dimensions of momentum times distance or energy times time. It was discovered by Max Planck in 1900 when he tried to explain the spectrum of radiation from a hot body (the "black body radiation problem").

proton: a positively charged particle found in the nucleus of all atoms.

quanta: the plural of "quantum."

quantum [from the Latin adjective meaning "how much"]: a fundamental unit, or "packet," of energy.

quantum theory: a theory of atomic and subatomic interaction based on the behavior of all particles as both waves *and* particles. The theory makes use of the concepts of probability and quantum energy to explain everything that happens on the atomic and subatomic scale.

uranium: a heavy, silvery-white radioactive element used in nuclear fuels, nuclear weapons, and research; the heaviest naturally occurring element (atomic number, 238).

X ray: a high-energy photon. X rays have a very short wavelength, and are emitted from a metal target when it is bombarded by energetic electrons.

FURTHER READING

Works by Niels Bohr

Bohr, Niels. *Atomic Physics and Human Knowledge.* New York: John Wiley, 1958.

———. *Atomic Theory and the Description of Nature.* Cambridge, England: Cambridge University Press, 1961.

———. *Collected Works,* vols. 1–10. Amsterdam: North-Holland, from 1972.

———. "Determination of the Surface-Tension of Water by the Method of Jet Vibration." *Philosophical Transactions of the Royal Society,* 209 (1909), pp. 281ff.

———. "Disintegration of Heavy Nuclei." *Nature,* 143 (1939), pp. 330ff.

———. *Essays 1958–1963 on Atomic Physics and Human Knowledge.* New York: Interscience, 1963.

———. "Neutron Capture and Nuclear Constitution." *Nature* 137 (1939), pp. 344ff.

———. "Resonance in Uranium and Thorium Disintegrations and the Phenomenon of Nuclear Fission." *Physical Review,* 56 (1939), pp. 418ff.

Bohr, Niels, and J. A. Wheeler. "The Mechanism of Nuclear Fission." *Physical Review,* 56 (1939), pp. 426ff.

Books About Niels Bohr

French, A. P., and P. J. Kennedy, (editors). *Niels Bohr: A Centenary Volume.* Cambridge, MA: Harvard University Press, 1985. As well done and fascinating as the Einstein Centenary volume, with many personal reminiscences of Bohr the scientist and Bohr the man.

Moore, Ruth. *Niels Bohr: The Man, His Science, and the World They Changed.* New York: Alfred A. Knopf, 1966. Straightforward and readable biography of Bohr.
Pais, Abraham. *Niels Bohr's Times: In Physics, Philosophy, and Polity.* New York: Oxford University Press, 1991. Somewhat complicated in structure and approach but overall the best and most detailed of the books on Bohr to date.
Rozental, S. (editor). *Niels Bohr: His Life and Work as Seen by His Friends and Colleagues.* New York: John Wiley, 1967. Highly personal reminiscences by those who knew him. Easy to read and often charming.

Books About Physics and Physicists of Bohr's Time

Boorse, Henry A., Lloyd Motz, and Jefferson Hane Weaver. *The Atomic Scientists: A Biographical History.* New York: John Wiley, 1989. Despite its slightly difficult structure and some occasional tough going there is much to be gleaned from this excellent book. Recommended for higher-level readers.
Cline, Barbara Lovett. *Men Who Made a New Physics.* Chicago: University of Chicago Press, 1987. Well structured and easy-to-read look at the scientists involved in the exciting early days of the quantum revolution. (Published in 1965 by Thomas C. Crowell and Co., New York, as *The Questioners.*)
Cassidy, David C. *Uncertainty: The Life and Times of Werner Heisenberg.* New York: W. H. Freeman and Co., 1992. A major biography of the still-controversial Heisenberg, probably best for higher-level readers.
Crease, Robert P., and Charles C. Mann. *The Second Creation: Makers of Revolution in 20th-Century Physics.* New York: Macmillan, 1986. Covers much of the same material as the Cline book, but in more detail and at a deeper level. Best for older readers.
French, A. P. *Einstein: A Centenary Volume.* Cambridge, MA: Harvard University Press, 1979. Wonderful selection of articles on Einstein the man and the scientist.
Frisch, Otto. *What Little I Remember.* Cambridge: Cambridge University Press, 1979. Reprinted Canto, 1991. A modest title for an

intriguing and readable little book by one of the participants in the history of the early quantum revolution.

Gamow, George. *The Great Physicists From Galileo to Einstein.* New York: Dover Publications, 1988. The great Gamow takes a look at some major historical physicists and their work. As always with Gamow, readable and enlightening.

———. *Thirty Years That Shook Physics,* New York: Doubleday, 1966. A highly readable classic, but now a little dated.

Han, M. Y. *The Probable Universe.* Blue Ridge Summit, PA: Tab Books 1993. Han follows up his successful book *The Secret Life of the Quanta* with a closer look at the mysteries and controversies as well as the theoretical and technological successes of quantum physics.

———. *The Secret Life of the Quanta.* Blue Ridge Summit, PA: Tab Books, 1990. Coming at the quanta from a practical rather than theoretical approach, Han concentrates on explaining how our knowledge of the quanta allows us to build such high-tech devices as computers, lasers, and the CAT scan. At the same time, he doesn't neglect either the theoretical or bizarre aspects of his subject

Heisenberg, Werner. *Encounters with Einstein.* Princeton, NJ: Princeton University Press, 1989. Intriguing firsthand look at one great scientist by another.

Hoffmann, Banish. *Albert Einstein: Creator and Rebel.* New York: Viking Press, 1972. One of the best, most reliable, and most readable studies of the thought and style of the great scientist.

———. *The Strange Story of the Quantum.* New York: Dover Publications, 1959. A highly readable classic, if a little dated today.

Jones, Roger S. *Physics for the Rest of Us.* Chicago: Contemporary Books, 1992. Informative, nonmathematical discussions of relativity and quantum theory told in a straightforward manner, along with discussions on how both affect our lives and philosophies. Very readable and recommended.

Jungk, Robert. *Brighter Than a Thousand Suns: A Personal History of the Atomic Scientists.* New York: Harcourt, Brace and Co., 1956. A not always reliable anecdotal account.

Lederman, Leon, with Dick Teresi. *The God Particle.* Boston: Houghton Mifflin Co., 1993. Highly readable and entertaining discussion of contemporary physics by the witty Nobel Prize-winning physicist.

McCormmach, Russell. *Night Thoughts of a Classical Physicist.* Cambridge, MA: Harvard University Press, 1982. A fascinating novel treating the mind and emotions of a physicist in Germany who attempts to understand the disturbing changes in physics and the world. It may be loosely based on Paul Ehrenfest, a friend of both Einstein and Bohr, who committed suicide in 1933.

Moore, Walter. *Schrödinger: Life and Thought.* Cambridge: Cambridge University Press, 1992. A major biography of the always intriguing and often controversial scientist, for higher-level readers.

Morris, Richard. *Dismantling the Universe.* New York: Simon and Schuster, 1983. A nontechnical and highly readable look at the thoughts and development that went into creating modern physics. As with other books by Morris, clear-headed and worthwhile.

———. *The Nature of Reality.* New York: McGraw-Hill Book Co., 1987. A clearly written and tough-minded look at some of the stranger implications of modern physics.

Pagels, Heinz R. *The Cosmic Code: Quantum Physics as the Language of Nature.* New York: Simon and Schuster, 1982. One of the best books on the subject, clearly and intriguingly written by the late Heinz Pagels, one of the most gifted of the modern writer-scientists.

Pais, Abraham. *Inward Bound: Of Matter and Forces in the Physical World.* Oxford: Clarendon Press, 1986. Thoughtful and insightful but a little on the difficult side.

———. *Subtle Is the Lord: The Science and Life of Albert Einstein.* New York: Oxford University Press, 1982. Occasionally tough going, but an excellent look at Einstein's thought and work.

Peat, David F. *Einstein's Moon: Bell's Theorem and the Curious Quest for Quantum Reality.* Chicago: Contemporary Books, 1990. Intriguing but occasionally difficult look at some of the stranger implications of the quantum world. Well worth sticking with through some of its tougher passages.

Ponomarev, L. I. *The Quantum Dice.* Philadelphia: Institute of Physics Publishing, 1993. A totally engrossing and well-conceived look at the history of atomic and quantum science by the Russian physicist. Well-balanced presentation of basic ideas and the scientists involved in their development.

Rhodes, Richard. *The Making of the Atomic Bomb.* New York: Simon and Schuster, 1986. One of the best and most thorough books on the subject by an always highly readable, entertaining, and well-informed author.

Sayen, Jamie. *Einstein in America.* New York: Crown, 1985. A good account of Einstein's life and thought focused primarily on his American years at the Institute for Advanced Studies at Princeton.

Snow, C. P. *The Physicists* (Boston: Little, Brown, 1981). Not always reliable factually, this book nonetheless conveys the spirit of the times written by a fellow physicist who was there.

Speyer, Edward. *Six Roads From Newton.* New York: John Wiley, 1994. Well conceived and straightforward examination of the great discoveries in physics from Newton to today. Nicely done and highly recommended for all readers.

Von Baeyer, Hans Christian. *Taming the Atom.* New York: Random House, 1992. Intriguing and informative. Von Baeyer concentrates primarily on the experimental side of the story, without neglecting the theoretical.

Weisskopf, Victor. *The Joy of Insight: Passions of a Physicist.* New York: New York: Basic Books, 1991. Charming, insightful, and easygoing recollections by a world-class physicist who studied with and knew the great makers of the quantum revolution.

INDEX

Illustrations are indicated by *italic* numbers.
The letter *g* after a number indicates a word in the glossary.

A

alpha particles (alpha rays) 21, 23, 27–30, 69–70, 106g
Andersen, Hans Christian 14, 15
Andrade, E. 26
argon 22
atom 9, 106g *see also specific topics (e.g., atomic models)*
atomic bomb 83, 84, *87,* 89, 90
Atomic Energy Commission, Danish 96
atomic mass (atomic weight) 9, 27, 106g
atomic models
    Bohr 38–45, *45, 50,* 71, 72
    Rutherford *24,* 29–30, 34, 69–71
    Thomson *see* "plum pudding/raisin-in-pound-cake" model
atomic number 27, 106g
atomic spectrum 39, 41, 106g *see also* spectral lines
atomic theory 9–11
atomic weight *see* atomic mass
Atoms for Peace Award 97
atom splitting *see* fission

B

Balmer, Johann 41, 43
barium platinocyanide 9
Becquerel, Henri 10, 21
Bell, John S. 75–76
beta particles (beta rays) 21, 28, 106g
black body radiation 35, 36
Bohr, Aage Niels (son) 48, 83–84, 86, 89, 90, *91,* 92
Bohr, Christian (father) 2, 3–5, 14
Bohr, Christian (son) 48, 78
Bohr, Ellen Adler (mother) 2–4, *3,* 77
Bohr, Erik (son) 48
Bohr, Ernest David (son) 48
Bohr, Hans Henrik (son) 48
Bohr, Harald (brother) *3, 7*
    childhood 2, 4
    death 98
    letters from Niels to 11–12, 14, 18, 29, 30
    letters to Niels 46

    letter to United Nations 92
    on Niels as lecturer 5–6
    at Niels's wedding 33
    sister Jenny's death 77
    university education 6, 11, 13
Bohr, Harald (son) 48
Bohr, Henrik George Christian (grandfather) 3
Bohr, Jenny (sister) 3, *3,* 4, 6, 77
Bohr, Margrethe Nørlund (wife) *12*
    Carlsberg House awarded to 73
    conversation with Pauli 57
    death of Niels 99
    engagement 14
    helps Niels write papers 14
    letters from Niels to 18, 19
    marriage as described by friends 13
    meets Niels 13
    meets Rutherford 33–34
    returns to Denmark after WWII 89
    travels with Niels 98, 99
    wedding 33
Bohr, Niels *7, 64, 91, 92, 103, 104*
    atomic bomb development *72,* 84–85, *86*
    atomic model developed by 38–45, *50,* 71, 72
    awards and honors 44, 73, 97
    Born's ideas accepted by 59
    childhood 2–5, *3*
    complementarity principle *see* complementarity principle
    death 99–100
    debate with Einstein xii–xv, 51–52, 60, 63–65, 75, 97–98, 100
    fission results received by 82–83
    at Institute for Theoretical Physics (Niels Bohr Institute) *see* Institute for Theoretical Physics
    last blackboard drawing *101*
    legacy of 103–105
    love of discussions 2
    meets future wife 13
    Oppenheimer quoted on importance of 66
    particle accelerators 72
    personal tragedies 77–78, 98
    quantum theory taught by 48

Institute for Theoretical Physics (Niels Bohr
  Institute) (Copenhagen, Denmark) 48–53,
  53, 56, 95, 96
isotopes 27, 107g

J

Joliot-Curie, Frédéric 71
Joliot-Curie, Irène 71

K

Kierkegaard, Søren 6
Klein, Oskar 70
Knudsen, Martin 34, 48
Kramers, Hendrik 70

L

Landau, Lev 70
Laue, Max 83
Lederman, Leon 59
Leucippus (Greek thinker) 9, 10, 106
Lie, Trygve 92, 93
Lorentz, Hendrik 9, 11
Los Alamos (New Mexico) 84–85
Lynghuset (Bohr summer home in Tisvilde,
  Denmark) 52

M

Manhattan Project 84, 85
Marshall, George 91
Maxwell, James Clerk 17
McGill University (Canada) 22
Meitner, Lise 81–83
Mendeleyev, Dmitry 9–10
models of atom see atomic models
Møller, Poul Martin 6
Mollgaard, Holgar 11
Morrison, Philip 76

N

Nagasaki (Japan) 89
Nazis 81, 82, 83
Nehru, Jawaharlal 98
neutrons 71, 107g
Newton, Isaac 17
New York Herald Tribune (newspaper) 93
New York Times (newspaper) 100
Nicholson, J. W. 39
Niels Bohr Institute see Institute for Theoreti-
  cal Physics
Nielsen, Jens Rud 34
Nobel Prize
  Bohr, Niels 44
  Einstein, Albert 37

Pauli, Wolfgang 57
Rutherford, Ernest 22
Thomson, J. J. 17
Nordita Institute (Copenhagen, Denmark) 95
Nørlund, Niels Erik 13
nuclear reactors 96
nucleus, atomic 22, 23, 30, 70–73

O

"On the Constitution of Atoms and Molecules"
  (Niels Bohr paper) 43
"On the Quantum Theory of Radiation and the
  Structure of the Atom" (Niels Bohr paper)
  47
Oppenheimer, J. Robert xi–xii, xv, 65, 84, 85
Ørsted, Hans 14

P

Pais, Abraham xii, xiii, 48, 91
particle accelerators 72, 95, 96–97 see also
  cyclotrons
particle physics 94
Pauli, Wolfgang xiii–xiv, 56–57, 62, 63, 70,
  73, 98
periodic table 10, 27–28, 28
Philip, Prince (England) 90
Philosophical Magazine 47
Philosophical Transactions (British Royal
  Society) 8
photoelectric effect 38
photons 42, 107g
Physics Society (Berlin, Germany) 51
Planck, Max 35–37, 46, 51, 62, 83
Planck's constant 36, 41, 42, 107g
"plum pudding/raisin-in-pound-cake" model
  (atom) 11, 23, 24
Podolsky, Boris 75
probabilistic physics 59
protons 70, 107g

Q

quantum 107g
quantum mechanics 44, 57, 59, 62, 105
quantum theory 36–39, 56, 60–61, 66, 75–76,
  105, 107g

R

radioactive displacement law 27
radioactivity 8, 10, 21–22
radium 21
Rayleigh, Lord (John William Strutt) 7, 8, 17,
  46
relativity, theory of 37
Röntgen, Wilhelm Conrad 8–9, 10, 21